STAIRWAY TO THE STARS

The Story of the World's Largest Observatory

STAIRWAY TO THE STARS

The Story of the World's Largest Observatory

Barry Parker

Drawings by
Lori Scoffield

PLENUM PRESS • NEW YORK AND LONDON

Library of Congress Cataloging-in-Publication Data

Parker, Barry R.
    Stairway to the stars : the story of the world's largest
  observatory / Barry Parker ; drawings by Lori Scoffield.
        p.   cm.
    Includes bibliographical references and index.
    ISBN 0-306-44763-0
    1. Astronomical observatories--Hawaii--Mauna Kea--History.
  2. Parker, Barry R. 3. Astronomers--United States--Biography.
  I. Title.
  QB82.U62M387   1994
  522'.19969'1--dc20   94-21016
                                                              CIP

ISBN 0-306-44763-0

© 1994 Barry Parker
Plenum Press is a Division of Plenum Publishing Corporation
233 Spring Street, New York, N.Y. 10013-1578

Printed in the United States of America

Preface

It is early morning. As I look out the window I see a fresh blanket of snow. The hedge, shrubs, trees, and lawn are all covered. The temperature is 12 degrees below zero. In the east the sun is rising over the distant hills, its rays reflecting from the windows of the houses across the street. It is quiet, with few cars passing this early in the morning. Overhead the sky is blue, fading to white near the horizon, with a light mist obscuring some of the distant objects. As I sip my coffee I think back to the beginning of this book. It is almost exactly a year ago that my wife and I set out for the Big Island of Hawaii. I was on a five-month sabbatical; during this time I would visit the observatories of Mauna Kea and spend a considerable amount of time at the headquarters buildings in Hilo and Waimea. I was also planning to fly to Honolulu to the Institute for Astronomy. The first of the giant Keck telescopes (Keck I) would come on line while I was there, and Keck II was being constructed. I was looking forward eagerly to the trip.

This book is a result of that sabbatical. In the first few chapters I cover the history of the observatory—the problems that were encountered and how they were overcome. One of my objects in writing the book is to give the reader a feeling for what astronomers do at observatories, so I have included chapters describing my visits to them both during the day and at night. And finally I have selected a number of astronomers and discussed their research in considerable detail. These discussions span black holes, cosmology, stars, the search for extraterrestrial planets, and the origin of our solar system.

It is difficult in a book such as this to avoid technical terms, but I have tried to limit them. For anyone unfamiliar with the few that I have used, I have provided a glossary at the back of the book.

I am particularly grateful to the scientists who assisted me. I talked to a large number of astronomers while visiting the observatories and headquarters buildings. In some cases they provided me with reprints and photographs. I would like to express my gratitude to all of them. They are Mitsuo Akiyama, Colin Aspin, C. Berthoud, Ann Boesgaard, Hans Boesgaard, David Bohlender, Len Cowie, Sandra Faber, Tom Geballe, Peter Gillingham, John Glaspey, Bill Heacox, George Herbig, Esther Hu, Bill Irace, Dave Jewitt, John Kormendy, Alan Kusunoki, Ron Laub, Olivier Le Fèvre, Bob McLaren, Terry Mast, Karen Meech, Guy Monnet, Kyoji Nariai, Jerry Nelson, François Rigaut, Ian Robson, Michael Rowan-Robinson, Dave Sanders, Barbara Schaefer, Gerald Smith, Malcolm Smith, Walter Steiger, Brent Tully, Richard Wainscoat, and Peter Wizinowich.

I would particularly like to thank Kevin Krisciunas for supplying me with a considerable amount of historical material, several photographs, for critically reading several of the early chapters, and for acting as a guide to several of the observatories. I would also like to thank Andy Perala for a guided tour of the Keck Observatories and several photographs.

Special thanks go to Don and Edith Worsencroft, Jack and Ann Roney, and Virginia Spencer for their hospitality while we were in Hawaii.

The line drawings were done by Lori Scoffield. I would like to thank her for an excellent job. I would also like to thank my editor Linda Greenspan Regan and the staff of Plenum for their assistance in bringing this book to its final form. And finally I would like to thank my wife for her support while the book was being written.

All photographs are by the author unless otherwise specified.

Contents

vii

CHAPTER 1

Introduction

Perched on the summit of the highest mountain in the Pacific, the Mauna Kea Observatory on the Big Island of Hawaii has a night sky of breathtaking beauty, a sky that gives the best astronomical observing conditions in the world. Not only is the sky dark, the air steady and incredibly transparent, but on the average at least half of the days are clear, with another quarter partially clear. No other observatory in the northern hemisphere approaches this.

Mauna Kea is one of two large volcanoes that dominate the Big Island. From the ocean floor it soars 32,000 feet to its summit, 13,796 feet of which is above sea level. It is so high that, even in tropical Hawaii, its top is frequently capped with snow in the winter. Although its neighbor, Mauna Loa, still spews lava down its sides every few years, Mauna Kea is dormant, its last eruption occurring about 3000 years ago.

Geographically speaking the Big Island is young, with Mauna Kea and Mauna Loa rising above the ocean less than a million years ago. The entire island is made up of lava that has flown from these volcanoes and others. In all, five volcanoes make up the island; besides Mauna Loa, only one other one—Kilauea—is active. It is sometimes called the "drive-in volcano" because it is the most accessible volcano in the world, with a road completely around its caldera (crater at the summit). Kilauea last erupted in 1983, and as of 1993 lava was still flowing out of it.

The Hawaiian islands were first populated by Polynesians from the south, probably Tahiti. A thorough knowledge of the

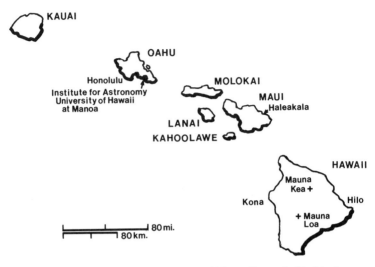

The Hawaiian Islands showing location of Mauna Kea on the Big Island.

stars—their rising and setting points along the horizon, in particular—made it possible for the Polynesians to navigate thousands of miles across the ocean. Religion played a large part in their culture, and with the islands being volcanic, it was perhaps inevitable that their gods would be associated with the volcanoes. Many Hawaiians, even today, still worship the volcano goddess Pele, whose home is the volcanic cavity of Kilauea.

The first white men arrived in 1778 when Captain Cook and his crew threw anchor on the shores of the Big Island. They were welcomed enthusiastically, but on a second visit a year later, fighting broke out and Cook was killed. Within a few years whaling boats began visiting the island, then came missionaries, and later entrepreneurs and adventurers.

The first ascent of Mauna Kea by a white man was made in 1823 when a Yale-educated missionary from Connecticut, Joseph Goodrich, scaled the mountain. According to his diary Goodrich began his final climb to the summit near midnight. After resting in

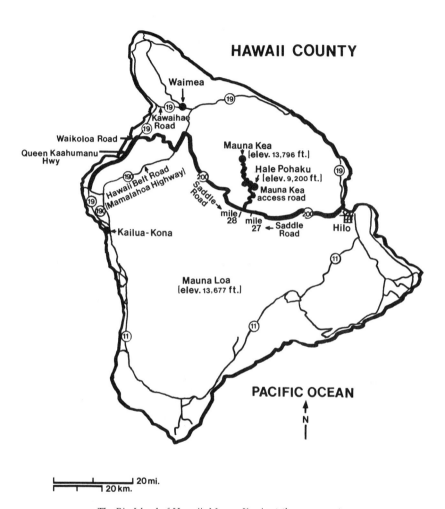

The Big Island of Hawaii. Mauna Kea is at the upper center.

Mauna Kea as seen from Hilo.

the afternoon, he rose about 11:00 P.M. to a full moon and made his way to the top, reaching it about 1:00 A.M. He found patches of snow and a small pile of rocks on the highest peak, probably left there by natives; the temperature was 27° F. He found the area to be extremely desolate, and devoid of vegetation. The top was studded with cinder cones rising out of fields of lava, pumice, and ash, but the absence of a large caldera indicated it was an old volcano. The air at the top was so thin it gave him a headache and "stomach sickness."

The famous botanist David Douglas (the Douglas fir is named after him) scaled the peak in January 1831. With a 60-pound pack on his back he found the final ascent extremely fatiguing. He visited several peaks in the region, and, like Goodrich, he experienced a violent headache. While on the summit he wrote, "Man feels himself as nothing—as if standing on the verge of another world. [There is] a death-like stillness about the place, not an animal nor an insect to be seen...."

Although neither man complained of inclement weather at the top (other than the cold) we now know that conditions can be severe. Blinding storms with winds over 100 miles an hour can occur during the winter. But when it is clear, the view of the sky is breathtaking, the stars like diamond chips spread over a sheet of black velvet, looking deceptively serene. Douglas wrote in his diary, "The stars shone with an intense brilliance." But it is unlikely that he could have imagined that one day this mountain would house the observatories we see there today. Not only is it the largest observatory complex in the world, containing more light-gathering power in its telescopes than any other observatory, but it also has the world's largest optical telescope—the Keck. Furthermore, it is the highest major observatory in the world, giving it access to radiation (e.g., infrared) that few other observatories are able to observe.

The story of the observatory begins in 1963 when Gerard Kuiper, the director of the Lunar and Planetary Laboratory at the University of Arizona, and former director of Yerkes Observatory in Wisconsin, came to Hawaii to look it over as a possible site for

Night at Mauna Kea. A time exposure showing star trails and car lights. (Courtesy Richard Wainscoat)

Stars over Mauna Kea. (Courtesy National Optical Astronomy Observatories)

an observatory. He was impressed with the observing conditions on the summit of Mauna Kea, referring to it as a "jewel." He approached the University of Hawaii with an offer to develop it jointly with them. John Jefferies, the head of the astronomy department at the University of Hawaii was receptive at first, but after pondering the details of Kuiper's plan he began having second thoughts. After considerable discussion the University of Hawaii decided to develop the mountain on their own. With a grant from NASA they built an 88-inch telescope, then encouraged others to locate their telescopes on the mountain, with the understanding that the University of Hawaii would get 15% of their observing time.

Several countries took an interest in the mountain, and in 1979 three telescopes were dedicated: a 3.6-meter telescope (a joint project of Canada, France, and Hawaii), a 3.8-meter United Kingdom infrared telescope, and a 3-meter NASA infrared telescope. With the completion of these instruments Mauna Kea became a world-class observatory, with more light-gathering surface area in its telescopes than any other observatory. In 1987 two more telescopes, sensitive to the submillimeter region of the spectrum (radiations of wavelengths slightly less than one millimeter), were added. The major coup for the University of Hawaii came, however, when the University of California decided to locate a 10-meter segmented mirror telescope on the mountain. It would be the world's largest optical telescope when completed. The project was funded in 1985 by the Keck Foundation through Caltech, and it became a joint Caltech–University of California project.

The Keck telescope was dedicated in November 1991. Earlier, the Keck Foundation had announced that it would fund a second, identical 10-meter telescope (now called Keck II) that would be placed 93 yards from the first Keck telescope. The two telescopes will be used together via a technique called interferometry; each will be part of a much larger telescope.

As construction began on Keck II, the Japanese broke ground for a telescope nearby. Called Subaru, after the Japanese word for Pleiades, a cluster of stars in the constellation Taurus, it will have

The Keck dome.

an 8.2-meter mirror that is sensitive to both visible and infrared. Unlike Keck, however, it will be solid, but thin—only 20 centimeters thick—and it will be supported with rods in back that will allow its shape to be adjusted slightly. The project is scheduled for completion in 1999.

Another large project called Gemini is still in the planning stages. A consortium of astronomers from the United States, England, Canada, Chile, Argentina, and Brazil is building two 8-meter telescopes, one in South America and the other on Mauna

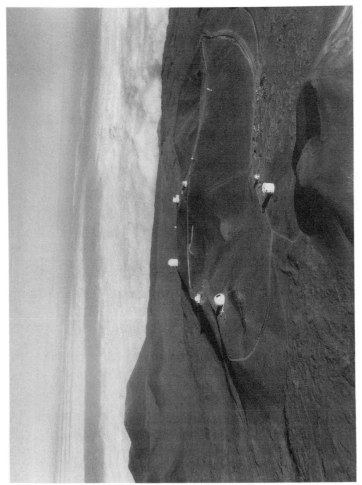

An aerial view of the domes on Mauna Kea. (Courtesy University of Hawaii Institute for Astronomy)

Several of the domes on Mauna Kea. The Caltech submillimeter dome is in the foreground.

Kea. A solid mirror will be used, and observations will be made in both the visible and infrared. It is scheduled for completion in 1998.

Smithsonian Astrophysical Observatory is planning to build a large submillimeter array on the mountain, consisting of six 20-foot dishes that will be movable. And finally, one leg of the worldwide Very Long Baseline Array (VLBA) was recently completed near the summit. Used along with several other radio dishes it will give a radio telescope with an effective diameter equivalent to the Earth's diameter.

The research that is being done with the telescopes on Mauna Kea is diverse, with hundreds of astronomers from many countries using them for thousands of different projects. Astronomers look forward eagerly to their night on the mountain. When the sun sinks below the horizon tiny points of light appear in the gradually darkening sky. As the sky blackens they blossom into a majestic panorama, a sight that would amaze most city dwellers. The Milky Way arcs overhead from horizon to horizon, cut here and there by dark clouds of dust and gas. Stars of all colors, e.g., red, blue, yellow—some solitary, others in pairs and groups—dot the sky, each etched against the blazing bulge of our galaxy.

As the sky darkens astronomers make last-minute checks and calibrations, then with the press of a button huge telescopes glide smoothly to preassigned positions in the sky and the night's work begins.

Stars, galaxies, clusters, and the like—anything that gives off light—are of interest to the astronomers on Mauna Kea. But lately considerable interest has also been directed at objects that give off no light—black holes. Late in their life, stars run out of fuel and the nuclear furnace at their center flickers and dies. The cooling core is no longer able to sustain the tremendous inward force of gravity and the star is suddenly overcome. In less than the blink of an eye the star collapses inward, crushing its atoms to incredible densities. Everything is squeezed to an infinitesimal point—a singularity—that is cut off from us by a surface called the event horizon. From a distance this event horizon is a bizarre-looking black

Stars and gaseous nebulae within the Milky Way galaxy (interior of the Rosette nebula).
(Courtesy National Optical Astronomy Observatories)

sphere, seen only because it blocks off background stars. It is tiny compared to the original star—only a few miles across—but it has properties that boggle the mind. Step inside—through the event horizon—and you are cut off forever from the universe. If you try to escape, you soon find it to be impossible; you are inexorably drawn toward the abyss at the center—the singularity.

For years it was assumed that black holes could only be created in the cataclysmic collapse of a giant star, but in the mid-1970s Stephen Hawking of Cambridge University of England showed that they could also be created in the big bang explosion, the explosion that created the universe. But these black holes would be different from those created in the collapse of a star. Tiny ones—smaller than an atom—would be possible; in fact, a whole range of black holes from small ones up to ones with a mass of a billion suns would be possible. Some of these massive black holes may have acted as "seeds" around which galaxies formed, and may now be at their core. If so, stars and gas would be whirling around them in a doughnutlike ring—an accretion disk—spiraling ever closer. As this matter spirals inward it is compressed and heated until finally it is so hot that x rays and gamma rays flood out into the universe. In some cases the matter pours into the black hole so fast it cannot be swallowed, and the excess is shot out through two tiny orifices, like toothpaste from a tube, giving rise to two violent cosmic jets.

The evidence for massive black holes in galaxies is now over-whelming. Astronomers have shown, in fact, that the strongest evidence comes not from violent galaxies far out in the universe but from nearby galaxies.

Although there is considerable work being done on nearby stellar objects, some of the effort on Mauna Kea is directed beyond the stars, even beyond the galaxies, to the depths of space. Our universe is made up not only of galaxies but of groups of galaxies, or clusters, and even clusters of clusters, or superclusters. Three-dimensional maps have been made of this structure revealing an unexpected texture on the largest scales. Clusters of galaxies appear to be stretched out in chains over the surface of huge voids—

The Whirlpool galaxy. (Courtesy Canada–France–Hawaii Corporation)

giant bubbles of nothingness. On this scale the universe appears frothy, like the soap bubbles in a bubble bath.

But within this seeming chaos is order: all of the galaxies are moving away from one another as a result of the expansion of the universe. Interestingly, though, astronomers have discovered that galaxies and clusters are not moving away from one another at the same rate. They are influenced by the galaxies around them, and therefore have two distinct velocities: a velocity related to the expansion of the universe (the Hubble flow), and a peculiar velocity that is superimposed on the Hubble flow as a result of the attraction of other galaxies or clusters. Both of these velocities are being studied at Mauna Kea.

As we look at the galaxies around us, we see that most are docile galaxies like our own, but in the outer regions of the universe we find different kinds of galaxies: exploding galaxies, blazars, and quasars. The outer reaches of our universe are infinitely intriguing; the galaxies are young and active, and we are seeing the universe as it was shortly after its birth. In theory we should be able to look all of the way back to its birth. Although we have not been able to do this yet, we are finding objects that formed only a billion years after the universe began (15 billion years ago).

With the large number of galaxies, and large number of stars in our galaxy, it seems reasonable that there might be life somewhere among them. Searches for planets that might support life have been made, but so far none have been found. When the Keck telescopes are in operation, however, a program called TOPS (Toward Other Planetary Systems) will be initiated, and an extensive search of all nearby stars will be made. And if there is life near any of them we should be able to identify it.

In this chapter we have had a brief glimpse into the history of the Mauna Kea Observatory and the research that is going on there. In the following chapters we will look into the struggles and difficulties that occurred along the way. It is an intriguing story—a story of frustrations and setbacks, along with successes and triumphs. We will also consider the research in much more detail.

CHAPTER 2

The Early Years

The first observatory in Hawaii was built shortly after the appearance of Halley's comet in 1910. It was a small building in a suburb of Honolulu that was equipped with a 6-inch refractor and a variety of other equipment. The College of Hawaii, which would later become the University of Hawaii, eventually took it over and operated it.

A PROMISING BEGINNING

In 1953 the small two-man physics department at the University of Hawaii in Honolulu became a three-man department when Walter Steiger arrived from the mainland. Known to his friends as "Walt," Steiger had fallen in love with Hawaii while stationed there during World War II. He knew it was the place for him, and he made a promise to himself that he would return. He had two years toward his degree at MIT before the war, and when the war was over he completed it. And as he had promised himself, he returned to Hawaii. To his disappointment, though, he found that despite his degree, there were no jobs. He wanted to teach but didn't qualify for public school teaching because he lacked the proper teaching credits. As a last resort he went to the physics department at the university, where, to his delight, he found that they needed a graduate assistant. It wasn't a full-time

Walter Steiger (in 1993).

job but it would support him, and over the next two years he obtained a master's degree in physics.

During this time he began asking himself what he really wanted to do with his life. He decided that he would like to teach at the University of Hawaii, but to do that he would have to get a Ph.D. While working on his master's degree he had taken a course given by a visiting professor from the University of Cincinnati. "He was a tremendous teacher," said Steiger, "the best I ever had. I told myself that if he's any indication of what the teachers at the University of Cincinnati are like I'm going there." And he did.

At the University of Cincinnati he worked under Boris Podolsky, who had earlier collaborated with Einstein on the famous Einstein–Podolsky–Rosen paradox. Steiger received a Ph.D. in 1953 and within a short time he got a letter from the chairman of the physics department at the University of Hawaii offering him a job. He was overwhelmed; it was the fulfillment of his dream.

Until now Steiger had never taken an astronomy class, and he had little interest in astronomy, but astronomy was soon going to play a large role in his life. Once in Hawaii he began thinking seriously about the type of research he would like to do. Little or no research was being carried out at that time, and Steiger wanted to change this. It seemed logical, he thought, to do something that took advantage of the natural environment of Hawaii—perhaps its mountains. Astronomy soon came to mind. But it wasn't night-time astronomy that attracted his attention; it was the study of the sun—solar astronomy.

Steiger was soon department head, and his dream was to set up a solar observatory. There were three high mountains on the islands: Haleakala (House of the Sun) on Maui, and Mauna Loa and Mauna Kea on the island of Hawaii. The latter two were out of the question; Mauna Loa was still active and Mauna Kea was remote and had no road to the summit. Haleakala was therefore selected, and in 1955 Steiger and a student, John Little, began testing conditions on the summit. Sky transparency was critical, but the number of clear days per month was also important. Within a short time they found that Haleakala was an excellent site, the atmospheric transparency even better than that at the Climax Solar Observatory in Colorado.

Steiger looked around for funds to build an observatory, but nothing was available. A breakthrough finally came in 1957. Earlier, July 1957 through July 1958 had been designated as International Geophysical Year (IGY). A large number of geophysical observations would be made throughout the world during this time, and Hawaii, being in the middle of the Pacific Ocean, was critical to this work. The first funds, therefore, were provided by the IGY, and with them a solar flare telescope was set up, not on Haleakala as Steiger had hoped, but on Oahu. Observers were to observe and record the times of flares occurring on the sun. It was a modest beginning, but the door had been opened.

Much of the excitement surrounding the IGY centered on the launching of a satellite. A network was needed to track the satellite and Hawaii was again critical, being one of the few stations in the Pacific. Steiger was called on to organize a team of observers. More important, though, was the establishment of a satellite tracking station, equipped with a telescope. Fred Whipple of the Smithsonian Astrophysical Observatory in Massachusetts was responsible for organizing part of the network. He got in touch with Dr. Kenneth Mees, the developer of Kodak Kodachrome color film, who was then in Hawaii, and asked if he would help. Mees was well-known to astronomers for his strong support; he had worked hard to develop specialized film for astronomers. Mees came to Steiger with an offer: he would donate $15,000 of Kodak stock for the building of a satellite tracking station if Steiger would oversee the land purchase and construction. Steiger was delighted. With the money he was able to build and equip a small station on Haleakala. The tracking telescope went into operation in July, 1957, and soon there were three full-time observers on the mountain.

Steiger's goal was still to establish a solar observatory at the site. The establishment of a tracking station was important in that it attracted attention to the site. Furthermore, in early 1961 Franklin Roach of the National Bureau of Standards came over to initiate studies of airglow and the zodiacal light. Collaborating with Steiger, he set up an extensive observing program. Finally, the university began to act. It decided to establish an Institute of Geophysics, distinct from the academic departments of the university. Faculty from physics, chemistry, geology, and other disciplines would make up the institute; they would teach part time for the university and do research the rest of the time in the Institute of Geophysics.

In late 1961 the National Science Foundation provided funds

for the establishment of the institute, and for the establishment of a solar observatory on Haleakala. Groundbreaking for the observatory took place on February 10, 1962. Over the next few months a 30-foot dome was built, along with dormitory space, kitchen facilities, offices, labs, and a machine shop. The telescope and support equipment were installed the following year.

The new observatory was dedicated in January 1964. It was named for Kenneth Mees, who had passed away in the meantime. With the establishment of a solar observatory the university was in a position to attract new faculty and Steiger hired three solar astronomers: Frank Orral, Jack Zirker, and John Jefferies. They were followed a little later by Marie McCabe.

While the solar observatory was being built, other important events were taking place. Gerard Kuiper at the Lunar and Planetary Laboratory of the University of Arizona had initiated a wide-ranging program to look for promising new observatory sites. Kuiper had come to the United States from the Netherlands in 1933 with a fresh Ph.D. from the University of Leiden. His first stop was the University of Chicago where he stayed for a number of years, later becoming the director of the Yerkes Observatory in Wisconsin. In 1960 he was lured away to the University of Arizona where he set up the Lunar and Planetary laboratory. Within a short time the University of Arizona had one of the best astronomy faculties in the United States.

Unlike most astronomers who specialize in stars and galaxies, Kuiper's passion was the planets and moons. The solar system was his domain, and he eventually became known as the "father of modern planetary astronomy." Among his discoveries was the detection of carbon dioxide in the atmosphere of Mars. He also showed that the atmosphere of Titan, Saturn's largest moon, contained both methane and ammonia. And he discovered satellites orbiting Uranus and Neptune. One of his greatest contributions, however, was his theory of the origin of the solar system. In 1951 he advanced the idea that the planets were formed when gases condensed to produce "protoplanets." His ideas are still central to the present-day theory of the origin of the solar system.

Gerard Kuiper is second from the left. Photo taken in the early 1960s. (Courtesy University of Arizona)

Sites at high altitudes were of particular importance to Kuiper. His fascination with the planets had sparked his interest in the infrared (heat radiation). Planets emit much of their radiation in the infrared, but infrared is absorbed by the moisture in our atmosphere. A high site, above most of the atmosphere, is therefore essential for infrared observations.

In his search for new sites, Kuiper had already scouted mountain peaks in California, New Mexico, and Chile. Because of their isolation and height, the mountains of the Hawaiian islands seemed promising, and in 1963 he came to check them out. Steiger took him up to Haleakala to the solar observatory. Kuiper was impressed and brought his assistant Alika Herring from the mainland to test the nighttime seeing at the site. Herring, who was part Hawaiian, had come to Kuiper's attention in 1960 after a number of his articles appeared in *Sky and Telescope* and other magazines. Kuiper was impressed with his skill and hired him. Besides being an excellent observer, Herring was an accomplished telescope builder; over the previous few years he had ground and figured 3500 telescope mirrors. The one he used on Haleakala, and later on Mauna Kea, was his best.

How good was the nighttime "seeing" on Haleakala? Steiger was sure it would be good, and within a short time Herring found it to be excellent. With continued testing, however, a problem surfaced. Haleakala was 10,000 feet high, very high as observatories go, but barely above the inversion layer on Maui. Dense clouds lay just below the peak, and they occasionally rose up to the solar observatory. Even worse, the observatory was on the edge of the Haleakala caldera, which frequently filled with fog, and sometimes in the evening wind would blow the clouds out across the observatory. It was soon clear to Herring that although the seeing and air transparency were excellent, Haleakala was not an ideal place for a nighttime observatory. But even as Haleakala fogged in, Herring could look across and see the peak of Mauna Kea.

About this time a letter appeared on Kuiper's desk in Arizona. It was an invitation from Mitsuo Akiyama of Hilo to consider Mauna Kea as an observatory site. The major event that led to this letter occurred on May 22, 1960; and it was an event that most people on the Big Island, and particularly the people of Hilo, would remember for the rest of their lives. The previous day an earthquake measuring 7.5 on the Richter scale had rocked Chile. Nine hours later another 7.5 quake occurred, then within seconds an earthquake 30 times more powerful struck. It measured 8.5 on the Richter scale.

A warning went out across the Pacific: Tsunami (large tidal

wave) possible. A network had been set up and it watched for a tidal wave. The wave from the first 7.5 quake struck Hilo the following day during the daylight hours. It was relatively small and did no damage. Residents in Hilo now waited for the second wave, created by the much larger quake. Traveling at almost 450 miles an hour it would take 15 hours to cross the ocean (6600 miles) and would arrive in Hawaii about midnight.

Residents were told to evacuate, but similar warnings in 1952 and 1957 had proved to be false alarms; in each case the wave was small. Furthermore the wave from the first quake had been small. Many people were complacent, and to make matters worse there was some confusion. Earlier the tidal wave alarm warning system had been changed from three sirens a few minutes apart to a single siren.

As midnight approached a number of people actually went down to the bay area to watch the wave as it came in. Just after midnight it struck. It wasn't a single wave, but three separate ones about 15 minutes apart, the last one being the most devastating. In places it was 30 feet high. According to observers there was a dull rumble just before it arrived, like a train in the distance. The wave grew steadily as it crossed the shallow bay. Then suddenly cars and houses alike were lifted in the air and carried inland. The force of the wave was so great it flattened steel parking meters throughout the bay area. Brilliant blue-white flashes occurred in many places in the city as the wave struck the power station.

The damage was overwhelming. The downtown area of Hilo was razed, literally flattened. And with it went hundreds of businesses and jobs. Over the next couple of years the Hawaii Island Chamber of Commerce looked around for something to pump up the economy. One of the first things to come to mind was the lava—the island was composed of lava. Maybe they could market it. Present at the meetings where such things were discussed was Mitsuo Akiyama, the executive secretary. One of Mitsuo's friends was Howard Ellis, the head of the weather station partway up the slope of Mauna Loa. Ellis would occasionally stop in Akiyama's office when he came down off the mountain. One day he said to him, "So often I look across at Mauna Kea high above the clouds and think it would be an ideal place for an astronomical observatory." Although he knew almost nothing about astronomy, Akiyama was intrigued; astronomy would bring in money—it would help the economy. There was, however, a serious problem: no road to the summit. They talked about the idea and decided to look into it.

In June 1963, Akiyama sent a letter to all large universities in the United States and the University of Tokyo. He got only one reply—from Gerard Kuiper. In the letter Kuiper said he would fly over in August and discuss the possibility of setting up an observatory. The meeting was postponed until November, then to December; finally, on January 12, 1964, Kuiper arrived for a five-day visit. The first couple of days he talked to various officials in Hilo, but his major interest was, of course, the summit of Mauna Kea, and on the third day he hired a pilot and spent several hours flying over the peak. He then wanted to get as close to it as possible from the ground, so a jeep ride was arranged for him to a ranger station called Hale Pohaku, partway up the slope, at an elevation of about 9200 feet. The road up to the summit would obviously have to begin at Hale Pohaku. When he returned from the trip Kuiper studied maps of the region for hours, planning a route. The land here was state-owned, and he would need permission from the governor. Akiyama therefore arranged a visit to Honolulu where Kuiper met Governor Burns and asked about a road. Burns was quick to see that an observatory would give an economic boost to the Big Island. Since NASA did not provide money for roads, Burns offered $25,000 for the building of it. Later, he added another $17,000.

While in Honolulu, Kuiper visited President Hamilton of the University of Hawaii, and George Woollard, the director of the Institute of Geophysics. He left for Arizona the next day. Earlier Kuiper had also made arrangements for Lyman Nichols, the District Wildlife Biologist, to hike to the top of Mauna Kea. Nichols made the trip in early February and took several slides. In his letter to Kuiper he gave a detailed description of the area. On the basis

The ranger station at Hale Pohaku. Photo taken at about the time of Kuiper's visit. (Courtesy Special Collections, University of Arizona Library)

of Nichols's photographs and the flight over the summit, Kuiper decided that the cinder cone Pu'u Poli'ahu, the second highest peak in the region, was the best place for the first telescope. The highest point was Hawaiian holy ground, and Kuiper decided it was best to avoid it.

In April, work on the road finally began. A route had been mapped out, but Kuiper was still unsure where he wanted the road near the top. In late April, with the road completed to the summit area, Kuiper returned and went with the bulldozer operator to survey the top and decide on the best route to the top of Pu'u Poli'ahu.

Everything was now ready for the dome and telescope. NASA had provided funds for a 12½-inch telescope and a dome, but the construction and installation were to be looked after by

Kuiper. The dome arrived in Hilo in early June and was taken up the steep road to the summit in sections. Water from a lake near the summit, Lake Waiau, was used to make cement, and over a period of 11 days the foundation of the dome was constructed. Akiyama arranged for several of his friends to help. With the completion of the dome, Alika Herring mounted the 12½-inch telescope and readied it for observations. Before observations could begin, however, arrangements had to be made for any emergencies that might occur. Storms on the mountain were severe, which posed a danger to anyone stationed there. A communication system was therefore arranged for Herring; he would have radio contact at all times with an amateur radio operator in Hilo by the name of William Seymour.

The dome of the 12½-inch telescope used by Alika Herring. It is no longer on the mountain. (Courtesy University of Arizona)

Gerard Kuiper is seated on the left. Mitsuo Akiyama is directly behind him. Howard Ellis is on the right. Photo was taken shortly after the dome was completed. (Courtesy Alika Herring)

Herring began his observations in the middle of June. Rather than use the mirror that came with the NASA telescope he used his own, a mirror acknowledged to be one of the best ever produced. It might seem strange, but Herring did not take photographs; instead, he made drawings of lunar and planetary features. It was soon obvious to him that the seeing was incredible, and the transparency of the atmosphere unmatched. Using a scale of 1 to 10, he classified each night (10 was perfect seeing). Many nights were 9 or better. Herring had observed at many sites around the world, and soon realized that he had never encountered a site superior to this. He was tremendously enthusiastic about it. After a few weeks he was relieved by a graduate student from the University of Arizona, William Hartmann.

On July 20 the observatory was dedicated. The ceremony was to have taken place at the observatory itself, but it was too windy, so it was changed to Hale Pohaku. About 200 people attended, including Governor Burns, Akiyama, Steiger, Woollard, and, of course, Kuiper. Several people gave speeches, including Governor Burns and Kuiper. In his speech Kuiper referred to the mountain as a jewel. "This mountaintop," he said "is probably the best site in the world . . . from which to study the moon, planets, and stars."

The mountain had its first telescope. It was small, no larger than many amateur telescopes, but Kuiper planned on shipping a 28-inch telescope and a 20-foot dome from Arizona as soon as an agreement with the University of Hawaii was made, then he would get money from NASA for a 60-inch telescope.

PROBLEMS

Kuiper assured officials at the University of Hawaii that he had money for a 60-inch telescope. He did, however, have some reservations about a joint venture with them. Steiger had mentioned to him that he was not getting as much support for his solar observatory on Maui as he would like. Although he was director of the observatory he had no secretary, no university vehicle to travel to the summit of Haleakala, and he was also concerned that the university had done little to try to protect Haleakala against man-made disturbances such as radio transmitters and excess lighting. Kuiper was worried that university officials would not push the project as hard as he felt it should be pushed. He wasn't sure, however, how to handle things. He didn't want to get off on the wrong foot immediately. The best thing, he decided, was to get a written agreement with them.

About this time Steiger stepped down as director of the Haleakala Observatory and appointed Jefferies his successor. Born in Australia and educated there and in England, Jefferies came to the Joint Institute for Laboratory Astrophysics in Boulder, Colorado, shortly after obtaining his Ph.D. Although he was a solar astro-

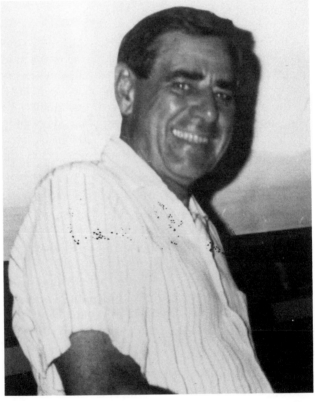

John Jefferies. Photo taken at about the time work on the mountain began. (Courtesy K. Kriscionas)

nomer, his specialty was theory, and he therefore had little experience with observing, nighttime observing in particular. He was, however, extremely ambitious and determined to take advantage of every opportunity that came his way. Because of this, many found him arrogant and egotistical. As one of his early students said, "He knew where he wanted to go, and if you got in his way, you got footprints up your back." He was quick to add, though, that he didn't appreciate the struggle that Jefferies and others were going through at the time. Many people, in fact, praised Jefferies for his strong leadership and firm hand in dealing with NASA and others.

Strangely enough, Jefferies was not enthusiastic about the project at first; he later admitted that his first ride to the summit of Mauna Kea was not an enjoyable experience. The road was dusty and he didn't tolerate the altitude well. When he got back down he said he hoped he would never have to go up again. Furthermore, he thought that the project was taking too much time away from his research. Within a short time, though, he had changed his mind.

Tests continued on the summit of Mauna Kea through the summer and into the fall. In all, Herring made four trips to Hawaii for observing runs; each lasted from two to six weeks. He was now sure that it was the best site he had ever seen.

In October 1964, Kuiper wrote to Woollard, the director of the Institute of Geophysics, trying to pin down an agreement between the University of Arizona and the University of Hawaii. He planned on bringing one of the world's most skilled observers at that time, Harold Johnson, to the new observatory. But uncertainty was beginning to cloud negotiations. Jefferies, Woollard, and others at the University of Hawaii were beginning to worry that an agreement with the University of Arizona would not be in their best interest. They didn't want a secondary housekeeping role, where they provided and maintained roads, while the research glories went to people at the University of Arizona.

Kuiper knew that the people at the University of Hawaii were hesitant, and he was cautious, but he wasn't prepared for the shock that came in late October. A letter arrived from Woollard informing him that the vice president of the university, Robert Hiatt, had visited someone in the astronomy section at NASA and inquired about the Mauna Kea project. Kuiper had implied that he already had money for the 60-inch telescope that was to be placed there. Hiatt was surprised to learn that although NASA was interested in developing Mauna Kea, no money had been promised to Kuiper. The real shocker, however, came when Kuiper read the latter part of the letter. Someone at NASA had told Hiatt that the best way for the University of Hawaii to get the site developed was to disassociate itself from Kuiper. They felt that Kuiper was overextended, and wouldn't be able to do the Mauna Kea project justice. Any proposal from him would likely be turned down. To make things worse, Woollard sent a copy of the letter to Governor Burns.

Kuiper was furious. He immediately went to NASA to find out who had been so outspoken. He talked to some of the senior officials, and they assured him that they had heard nothing about any prejudice against him, and were stiu extremely interested in proceeding with the Mauna Kea project. He was given no assurance, however, that the money would go to him.

Kuiper wrote back to Woollard trying to placate him. He told Woollard he had talked to senior officials and everything was okay. Someone in the astronomy section (he didn't mention any names) had just been a little outspoken.

Woollard and Jefferies were not convinced. Furthermore, by now it was known that Donald Menzel of Harvard had become interested in the site. It was Kuiper, in fact, who had drawn Menzel's attention to the site. Over a period of years it had become clear to Menzel that Harvard was losing valuable personnel because it didn't have an adequate telescope. Menzel was in the process of looking for a site.

Despite his determination, Kuiper was beginning to lose faith in the project. He wrote to Oran Nicks, the director of the lunar and planetary program at NASA, telling him that if NASA felt that he was creating a problem he would gladly withdraw from the project, as long as the site was properly developed by competent people. He was sure that the University of Hawaii wasn't capable. They didn't have a proper astronomy program—only four solar astronomers.

Kuiper was still hoping for some sort of an agreement with the University of Hawaii. But it didn't come. He soon got a letter from Hiatt informing him that the University of Hawaii had decided they didn't want to rush into anything, and at this point they weren't going to sign an agreement. In a joint venture with the University of Arizona, he said, Hawaii was going to have to put up a lot of money and they wanted to be sure they were doing the right thing. Since Harvard was now also interested in the site they also had the option of signing a contract with them. When they later saw Harvard's proposal, however, they rejected it immediately. Harvard was much less generous than the University of Arizona in giving them anything more than a maintenance role.

A third possibility, however, now existed. NASA was still undecided on who should get the money. They had accepted proposals from the University of Arizona and Harvard, but there was no reason why the University of Hawaii couldn't submit their own proposal. And within a short time they decided this was the best route for them.

Kuiper could hardly believe it when he heard that the University of Hawaii was submitting a proposal. It didn't have any night observers, and no one on their staff knew anything about the design and building of telescopes. It had next to nothing to offer as far as he was concerned. Still, the University of Hawaii did have one thing going for it: it was close to the observatory site, and running it would be no problem. The University of Arizona was thousands of miles away.

John Jefferies was now determined to get the project for the University of Hawaii. During January, 1965, he was hard at work gathering information and writing a proposal. After consulting with many people, Jefferies decided on an 84-inch telescope. (Both the University of Arizona and Harvard had proposed 60-inch telescopes.) Not only was it larger than the other telescopes, but it was also to be fully automated. Jefferies asked $3 million for the design and construction of the telescope. The State of Hawaii, through the university, would put up $2.5 million for the dome, other buildings, and a power line; they would also maintain the road and midlevel facilities. Jefferies hired Charles Jones of Los Angeles to design the telescope. Jones had earlier designed the 84-inch telescope at Kitt Peak Observatory. Jefferies also later hired Hans Boesgaard as chief engineer.

When Kuiper heard that Jefferies had hired Jones he shot off a letter to Woollard saying that Jones was extremely expensive. By this time Kuiper had become severely depressed; he had planned on sending Harold Johnson to the site, but Johnson had now become disillusioned because of the fighting and had gone off for several months to Mexico. Still, Kuiper was determined to get the project for the University of Arizona.

In February the University of Hawaii submitted their proposal. Strangely, it didn't specify Mauna Kea, but stated that both Mauna Kea and Haleakala would be thoroughly tested for seeing, transparency of atmosphere, and number of clear days, and the better of the two sites would be chosen. It also specified that the university would build up their astronomy program. Within a short time, in fact, the Institute for Astronomy was created with Jefferies as its director.

Upon hearing that the University of Hawaii's proposal was at NASA, Kuiper couldn't contain his curiosity. He wrote to the director of NASA asking to see a copy. At this point he was still sure there was no way that the University of Hawaii could follow through and produce a usable telescope. Nevertheless he wanted to get his hands on the proposal, and when he saw it he was impressed. They had obviously done their homework.

With the proposal off to NASA, Jefferies began sending observers to the summit of Mauna Kea to test the observing conditions (they already had a solar observatory at Haleakala so it was easy to test). Strangely, Herring was still there, performing tests for Kuiper, so there were now two groups on the mountain. Because of the danger from storms, Jefferies specified that there always be two observers on the summit from the University of Hawaii. Herring helped them set up their equipment.

By now even Herring was becoming disillusioned. He wasn't sure what was going on. Writing to Kuiper about this time he said, "There's a lot of dirty politics here." Akiyama had told him that the Chamber of Commerce of Maui was pressing hard for the project; they wanted it on Haleakala, and at one point while he was campaigning for Mauna Kea, he was told to mind his own business and stop interfering. "If the project went to Mauna Kea there would be a lot of explaining to do to the people of Maui," someone had told him. Herring wondered if there was any sense in doing further tests. He was even beginning to wonder if all of the work he had done had been worth it. Everything, he was sure, was going down the drain.

NASA now had three proposals to choose from, two from large universities and one from the relatively small University of Hawaii. To everyone's surprise they went out on a limb and selected the proposal from the University of Hawaii.

Kuiper couldn't believe it; he was furious. After all he had put into the project it was a serious blow, and he was bitter about it for years, telling friends and strangers alike that he had discovered Mauna Kea and it had been stolen from him. Soon, however, he was so busy with other projects, he began to forget. He died in Mexico in 1973, just three years after the completion of the University of Hawaii telescope.

Officials at the University of Hawaii were jubilant when they heard that their proposal had been accepted, but they now had a big job ahead of them. In addition to designing and building the telescope and dome, they had to build up their astronomy program. The Institute for Astronomy had been set up, but they now needed personnel. Jefferies wanted to hire planetary astronomers with experience—well-known ones with good reputations, if possible. But none was available so he had to settle for young, fresh graduates. Dale Cruikshank, David Morrison, Anne Boesgaard, Sidney Wolff, and William Sinton were among his first recruits; all later became leaders in the field.

THE BUILDING OF THE 88-INCH TELESCOPE

Once the testing was over it was evident that Mauna Kea was definitely superior to Haleakala, and despite his earlier reservations Jefferies settled on Mauna Kea. The first order of business was pouring the cement for the base of the dome, but work had barely begun when winter was upon them. Worst of all, it was one of the hardest winters on record. Blizzards, with winds of 100 miles an hour, raked the top. Snow piled up to unheard-of depths, and the ground was frozen for weeks on end. They had to use compressed steam to thaw it out. The wind was so fierce at times it would toss the heavy cement buckets around, making it dangerous for the workers to get close to them. Even worse than the harsh winter, though, was the altitude. Workers had never worked at such heights before. With oxygen roughly only half that at sea level, they exhausted easily, and the turnover was large.

Meanwhile the telescope arrived in Hilo where it lay waiting for the completion of the dome. It had been sprayed with a protective lacquer, and with the heat and long delay, it baked into the metal. It tooks weeks to get it off.

One piece of good luck did come their way, however. Jefferies ordered an 88-inch blank disk for the telescope mirror, hoping to get 84 usable inches. The disk was perfect and they got all 88 inches.

Other problems also plagued them. When the road to the summit was built, skiers began to take advantage of it to get to the snow at the top, but when work on the telescope began the road was closed to the general public because of the possibility of vandalism. The skiers felt deprived of a right they considered to be theirs. Hunters soon joined in the opposition as they became concerned about the disturbance that traffic up the mountain was causing to the sheep and wild boar on the slopes below the observatory. For several years these groups along with environmentalists objected strenuously to some of the activities on the mountain. One thing they particularly frowned on was a planned expansion at Hale Pohaku in the early 1980s. As the observatory

The 88-inch telescope. (Courtesy University of Hawaii Institute for Astronomy)

grew it became increasingly obvious that a much better midlevel facility was needed to house astronomers during their observing runs. Because the observatories are at such a high altitude, it is essential that the observers do not return to a low altitude during their observing run. It has been found that they can easily acclimatize to the thin air at the top if, after observing all night, they spend their days sleeping at Hale Pohaku.

The 88-inch telescope was to be completed the summer after construction began in the fall of 1967, but because of the severe winter it was delayed, not just a year, but two years. It was finally dedicated and opened in June 1970. Kuiper was invited to the dedication, and much to everyone's surprise, he attended. Afterward he inspected the telescope and facilities and said he was impressed. Still, the problems were not over. For the next five years astronomers struggled to make the telescope's sophisticated mechanical and electrical systems work. The major problem, according to Bill Heacox, a graduate student at the institute in the early 1970s (he used the telescope for his thesis project), was the computer system that ran it. The 88-inch was one of the first telecopes in the world to be fully computer controlled. "A large part of the problem was the computer," said Heacox. "It never worked well, and it put off a tremendous amount of waste heat." Heacox also mentioned that the computer language that was originally used—FORTRAN—was not designed for tasks such as controlling a telescope. "Things were a real mess for a while," said Heacox.

In 1976 Tom McCord was brought in to completely redo the computer and electronic system, and it was soon working extremely well.

When the telescope went into operation in 1970 it was the eighth largest in the world.

CHAPTER 3

Expansion and New Telescopes

After work started on the 88-inch telescope, Jefferies turned some of his attention to expanding the Institute for Astronomy as well as trying to interest others in Mauna Kea. There was a considerable amount of room on the top of the mountain for other observatories, and with the exceptional conditions, he was convinced others would be interested. And indeed within a few years, he had attracted the attention of several countries, Canada and France among the first.

THE CANADA–FRANCE–HAWAII TELESCOPE

In the late 1960s, Canadian astronomers decided they needed a larger telescope. No major observatory had been built in Canada since the David Dunlap Observatory was constructed in 1935. They were able to convince government officials that the need was genuine, and plans were made for a telescope about 3 to 4 meters in diameter. A search was made throughout Canada for a suitable site, and a peak in the Okanogan Valley in central British Columbia was selected. Called Mount Kobau, it was close to a radio telescope that was located in the area.

A road was constructed to the summit of Mount Kobau and a testing program was initiated. Astronomers installed a 16-inch telescope in a small dome to check the seeing, transparency of atmosphere, darkness of the night sky, and the number of clear nights. As plans continued the observatory was given a name: Queen Elizabeth II. It proved to be a good site, with many cloudless nights during the summer, but the winters were a problem. Furthermore, although the seeing was the best in Canada, it was not excellent, and at only a few thousand feet above sea level, the site was not exceedingly high. This meant that very little infrared would penetrate to the summit, and infrared radiation was becoming increasingly important to astronomers. Because of these and other problems, doubts about the site began to arise. Astronomers began to ask themselves: Would it be better to place the telescope at a superior site outside Canada? Eventually, because of the uncertainty, the project was put on hold.

About the same time something similar happened in France. Astronomers had decided that they needed a new telescope, and the government had agreed to support it. A search for a site was made and a suitable peak was located in the south of France. As in the Canadian case, it was tested and eventually astronomers began to have second thoughts. Finally, the project was put on hold. John Jefferies, who was on sabbatical in France at the time, heard about the dilemma, and invited French astronomers to take a look at Mauna Kea. They visited the mountain and were impressed.

Soon after the French made their decision to locate on Mauna Kea, Graham Odgers of the Dominion Astrophysical Observatory in Canada was in France. He heard about the French project and talked to some of the people involved, telling them about the problems Canada was having. Since both countries were short on funds, he suggested that they work together. The French thought it was a good idea.

Many Canadian astronomers were disappointed at first. Hawaii was so far away, and they would have to share the telescope with France and the University of Hawaii. They soon discovered, though, that with the excellent weather on Mauna Kea, they would get just as many observing days as they would have with a telescope on Canadian soil. In addition the Hawaiian site had many other advantages: seeing and transparency were much bet-

The pier of the Canada–France–Hawaii telescope during construction. (Courtesy Canada–France–Hawaii Corporation)

ter than on Mount Kobau, and with Mauna Kea being so high, observations could be made in the infrared. There was also the advantage of skies free of light pollution; the only large city on the island is Hilo, and it is usually buried beneath the clouds associated with an inversion layer.

Once the Canadians and French had agreed to work together, they had a number of decisions to make. Both countries had drawn up plans for a telescope. Which of the two plans would they use? The two designs were similar, but the French design was more easily adapted to Hawaii's latitude, so, after some discussion, it was used. In addition, the Canadians had bought a glass blank for their telescope mirror; it was sitting in a box in a parking lot in Victoria. Would it be adequate?

The Canadian mirror had been cast under the assumption that the curve on its surface would be a hyperbola. With the new mount and site, however, a parabolic curve seemed more appropriate, and it would be difficult to parabolize the Canadian blank. Furthermore, a new material called Cer-Vit had come on the market. It was a ceramic product, similar to glass, that had an extremely low coefficient of expansion. A low coefficient of expansion is critical in a mirror of this type, as expansion or contraction creates distortion. After considerable discussion a new disk of Cer-Vit was purchased.

An official agreement between Canada, France, and the University of Hawaii was signed on October 25, 1973. A 3.6-meter (142-inch) telescope would be built at a cost of $30 million; Canada and France would each put up half, with the University of Hawaii providing the land and midlevel facilities. Canada and France would also each provide 45 percent of the operating costs, and the University of Hawaii, the remaining 10 percent. In return, the University of Hawaii would get 15 percent of the telescope time, with the rest divided equally between Canada and France.

The groundbreaking ceremony took place in July 1974. It was a promising start, but many problems arose before the observatory was complete. The first came shortly after construction began. The project manager, R. Waiguny, installed a trailer for the workers at

Dedication ceremony for the Canada–France–Hawaii telescope. (Courtesy Canada–France–Hawaii Corporation)

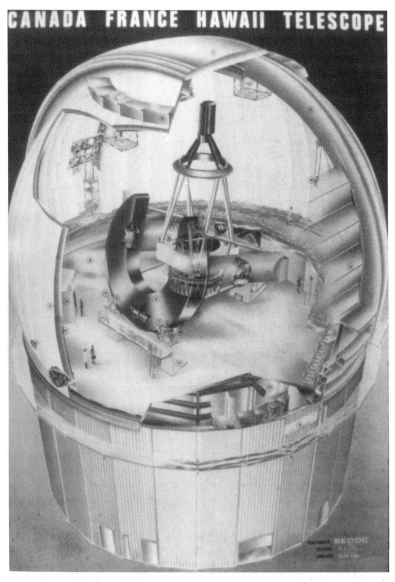

Cutaway of the dome showing the Canada–France–Hawaii telescope. (Courtesy Canada–France–Hawaii Corporation)

the midlevel facility, Hale Pohaku. According to county officials he didn't get a permit, and within a short time the Land Board Chairman began to threaten him. "If there is any further trouble, we will go to the Attorney General and stop the whole project," he said. He gave Waiguny only a few days to contact Canadian and French authorities; if the problem wasn't settled a fine of $500 a day would be levied.

"The early days were difficult," said C. Berthoud, who is now administrative manager of the Canada–France–Hawaii (CFHT) facility. "We had no office in Hawaii at first and that was a mistake. There was no spokesman here." Waiguny, the construction manager, acted as spokesman and he wasn't sure how to deal with some of the difficulties.

They no sooner had the trailer issue resolved when another controversy arose. Environmentalists, hunters, and a number of others began objecting to the development of the mountain. A "master plan" was called for.

Later there were difficulties when Canada and France began setting up their headquarters. The Institute for Astronomy wanted them to locate on the Manoa campus in Honolulu. And for a while a small office was set up there. They soon found, however, that they preferred to be on the Big Island, close to the telescope. But where? For a lot of people Hilo was the obvious choice; it was the largest city on the island, and the seat of government. A small temporary office was set up in Hilo in 1976.

About this time, however, CFHT officials were approached by a group from the Chamber of Commerce of Waimea, a small town in the northern part of the island. It was roughly the same distance from the top of Mauna Kea as Hilo, and it got less rain.

Waimea is home of the famous Parker Ranch, the second largest cattle ranch in the United States. Parker Ranch offered to lease land to CFHT offficials for their headquarters. Then the Hawaii Island Chamber of Commerce came to them with a presentation, encouraging them to build permanently in Hilo, and for a while it was a toss-up where the headquarters would be built. After looking at both offers carefully, CFHT officials decided to buy land in Waimea, and locate there.

The telescope was completed in 1979, and the dedication ceremony took place on September 29, 1979. "It was an exciting day," said Berthoud. "The governor of Hawaii, a high-level government minister from both Canada and France were here." The ceremony took place at the dome, with each of the top-level people giving a speech.

THE NASA INFRARED TELESCOPE

The CFH telescope was considered a tremendous breakthrough for Jefferies. He was determined to make the observatory one of the biggest and best in the world, and he had a good start with the CFH telescope. But the Canadians and French weren't the only ones interested in Mauna Kea. In the early 1970s NASA decided it needed a large infrared telescope to help support its planetary satellite program. A committee, chaired by Jesse Greenstein of the California Institute of Technology, decided that a 3-meter infrared telescope should be constructed. It was critical that it be placed at a high altitude where the infrared was accessible. James Westphal was commissioned to look for a site. He checked out mountains in the United States, Chile, and Mexico and decided that Mauna Kea was the best.

With the news that the telescope might be built on Mauna Kea, a conflict soon arose between the University of Hawaii and the University of Arizona. The University of Arizona had a strong infrared group that preferred to have the telescope as close to them as possible—preferably on Mount Lemmon in Arizona. They derided the small infrared group at the University of Hawaii and argued that Mauna Kea was too remote and working conditions far from ideal. They were also convinced that the seeing was no better than at Mount Lemmon. Part of their prejudice against the site was a result of a visit by two Arizona astronomers who came to conduct tests on the summit of Mauna Kea in the late 1960s. Ignoring warnings about acclimatizing, they went directly to the summit without staying over for an hour at Hale Pohaku. As a result they suffered altitude sickness, and to make things worse, an invisible ice storm passed across the summit while they were making their run and they got disastrous results. They left with a poor opinion of the mountain.

Something else that worked against the Institute for Astronomy was the 88-inch telescope. It had been completed in 1970, but for the first five years there were difficulties with the electronics, and the computer system that drove it. The people working on the telescope tried to keep the problems to themselves, but rumors eventually got out.

"When I was working on my thesis, someone scored the declination ring of the 88-inch telescope," said Bill Heacox, who is now director of the Space Science Department at the University of Hawaii in Hilo. "The tension . . . or something wasn't right and they ruined several teeth. We had to take it out and ship it back to the manufacturer, so the telescope was down for about three weeks. Nobody wanted to talk about it. Everything was hushed up until we got it back together. They didn't want anyone to know about it."

Heacox said that part of the problem with getting the NASA telescope was "that people just didn't want to believe how good the seeing was here. We couldn't seem to convince them we were giving them accurate data. There was a lot of suspicion we were cooking it." Heacox remembered a rumor that was floating around at the time: someone at the University of Arizona had said, "If the project goes to Mauna Kea, we should decide not to build it."

Despite the opposition, NASA decided to go with Mauna Kea, and in February 1974, a contract was awarded to the University of Hawaii. The infrared group at the University of Arizona was furious.

Jefferies went to work immediately. He hired Charles Jones of Los Angeles to design the telescope; Jones had earlier worked on the 88-inch telescope. And he hired a project manager.

Meanwhile a science advisory group had been set up by NASA to check on the progress of the work, and make recom-

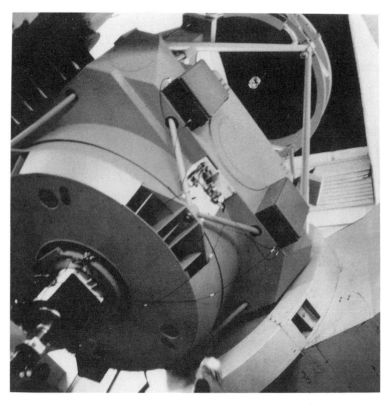

The NASA infrared telescope. (Courtesy University of Hawaii Institute for Astronomy)

mendations to them. Realizing that many of the infrared measurements would be made during the day the committee specified that the telescope mount be particularly rigid. During the day it is possible to detect some of the brighter stars and they could be used as guide stars for locating infrared sources that could not be seen. A given infrared source, however, might be as far away from a guide star as 15 degrees. The committee therefore recommended that the operator of the telescope should be able to move it 15 degrees from a given position to another position, with an accuracy of less than 2 seconds of arc. No other telescope in the world could do this. But for the infrared telescope, the committee thought it was essential.

Jones was convinced he could accomplish this with a "tuning fork" mount. The advisory committee, however, was skeptical. They preferred a "yoke" mount, which was known to be much more rigid. A tuning fork mount, as the name implies, looks like a tuning fork, with the telescope mounted between the prongs of the fork. It is a simple design and easy to drive equatorially (so that it compensates for the Earth's rotation). You merely have to point the handle of the tuning fork along the Earth's spin axis (approximately in the direction of the north star, Polaris) and drive it slowly around this direction. The yoke mount, on the other hand, consists of a large rigid horseshoelike structure with the telescope mounted in the center. The 200-inch Palomar telescope has this design, as does the CFH telescope.

The science advisory group recommended a yoke design, but Jones and Jefferies disregarded their advice and went ahead with a tuning fork design. Many people were unhappy. They were convinced the tuning fork design was too flexible, and would not meet the specifications. Jones insisted it would. The bickering continued into late 1975, and by then redesigning the telescope would cause a delay, so NASA decided to go with Jones's design.

Then tragedy struck. The 3-meter mirror had been sent to Kitt Peak National Observatory in Arizona for grinding and polishing. Before they could start, however, a hole had to be cut in the center of the mirror to allow light from the secondary mirror to pass through it to the instruments below. This type of arrangement is called a Cassegrainian. After drilling the hole, workmen left for the Labor Day weekend. When they returned the following Tuesday they discovered a crack radiating from the center hole outward across the mirror. They were shocked. Although this had been known to happen, it was rare, and was therefore completely unexpected. The mirror was ruined. Not only would they have to purchase a new mirror which would delay the project, but the first mirror had been one NASA had used in testing for the space telescope, and they had got it free. A new mirror would cost several hundred thousand dollars.

Some of the people at the University of Arizona, and in the advisory group, were now saying they had known from the beginning the project would fail. Furthermore, many of Jefferies's group were disheartened. It had been too much trouble. They were battling with their own telescope, trying to make it work better, and working with the CFH group. Jefferies, however, was stubborn. He had no intention of giving up.

With the cracking of the mirror there would be a considerable delay. NASA decided to use this time to take a second look at the project. Was the telescope really needed? Was Jones's design adequate? The advisory committee voted for a complete revamping of the plan and NASA went along with it. The upshot was that Jones was out, and the yoke design was in. The project manager was also let go, and Gerald Smith of the Jet Propulsion Labs in California was brought in to oversee the project.

Jefferies was furious, but he bit his tongue and said little. After all of the work he had put into the project he didn't want to lose it, so he went along with NASA. Smith arrived in January 1976, and, as it turned out, he was the ideal man for the job.

"There was a lot of dissension when I arrived," said Smith. "There was also some dissatisfaction with the project manager, who had a background in construction, and knew little about telescope making." Smith immediately went to the advisory group and discussed the project with them. "They told me they preferred a yoke design, and we soon agreed that the yoke mount was the way to go," said Smith. He then arranged for some JPL and Kitt Peak people to design a new telescope. This time a computer was used to determine the deflections at various positions of the telescope. "We modeled a yoke design and found that deflections were very low," said Smith. "Also, I had a strong background in planning, budgeting, scheduling and telescope making so we managed to convince the advisory committee quite quickly that the plan was feasible, and they advised NASA to go ahead with it."

The telescope, which is now know as the IRTF, was built in about 3½ years, and it came up to NASA's specifications. "Everything went smoothly after that," said Smith. "The telescope has been on line for 15 years now and it's still an excellent instrument."

THE UNITED KINGDOM INFRARED TELESCOPE

While the University of Hawaii was still in the initial throes of battling for the NASA infrared telescope, astronomers in Great Britain became interested in building a large infrared telescope. The original proposal came from Jim Ring of Imperial College, London, and the late Gordon Carpenter of the Royal Observatory in Edinburgh, Scotland. They hoped to get a telescope with a mirror approximately 4 meters in diameter, but money was in short supply, and they were told that if they asked for enough to do a proper job, they wouldn't get any. There was the suggestion, however, that if they asked for half what they really needed they might get it. So they asked for the smaller amount and got it. The problem then was how to get a 4-meter mirror and build a telescope with what they had. They would obviously have to cut some corners. The first thing they did was consider a mirror about half as thick as it should have been. This would save a lot, but it meant that the mirror would need substantial back support, if it wasn't to sag and distort as it was moved from position to position.

Design of a support cell began in 1973. Calculations suggested

The United Kingdom infrared telescope.

that if supported correctly, a thin mirror could approach the performance of a conventional thick one. A small prototype was built and tested in the Canary Islands. It worked well, so they ordered a thin mirror. It was supported by 80 pneumatically controlled pads arranged in three concentric circles; radial support was provided by 24 counterbalanced lever arms, bonded to the mirror's edge and pivoted at points attached to the cell. It turned out to be an extremely successful system.

The next thing they looked at was the dome. It was cut down in size until it was barely big enough for the telescope. If you

The dome of the United Kingdom infrared telescope.

A close-up showing the instruments at the base of the United Kingdom infrared telescope.

compare the dome of the CFH telescope (3.6 meters) to it, you see that, although the United Kingdom infrared telescope (UKIRT) mirror at 3.8 meters is larger in diameter, it is housed in a considerably smaller dome.

The dome and telescope were constructed in England. Both were completed at the end of 1977, and after dismantling, the telescope was shipped along with the mirror and other optics to Hawaii. By July 1978, everything had been assembled on the top of Mauna Kea, and the telescope was ready. To everyone's delight the original specifications were surpassed, and there was no serious cost overrun. For its size it was the least costly telescope on the mountain; its final cost was about $5 million.

The dedication came on October 10, 1979, when it was officially opened by the Duke of Gloucester. The low cost of the telescope turned out to be a mixed blessing. On the negative side, it didn't track well at first. "For the first two or three years it was quite wobbly," said Kevin Krisciunas, a software specialist who has worked with the telescope. "There was a joke around at the time that it would guide to within 30 arc seconds over a couple of hours, but it also guided to 30 arc seconds over a couple of minutes." There were also problems with the instrumentation. On the positive side, however, the thin mirror has turned out to have an advantage. Using the pneumatic supports behind the mirror, astronomers are able to warp and distort it, and this has proven to be particularly useful in relation to a number of modern techniques.

The telescope can now be operated remotely from Hilo, and from several centers in England. Edinburgh is considered to be the major headquarters for the telescope but it was felt necessary to have a significant presence in Hawaii. A small headquarters was therefore set up in Hilo shortly after construction on the dome began. For a while there were plans to locate next to the CFHT headquarters in Waimea, but negotiations broke down and the temporary office in Hilo was enlarged. In 1985 a permanent headquarters building was constructed.

PROBLEMS ALONG THE WAY

The year 1979 was an important one for John Jefferies and the Institute for Astronomy. Three major observatories were dedicated: the CFHT, the IRTF, and the UKIRT. Jefferies's dream was coming true. Mauna Kea was rapidly becoming a world-class observatory. The route, however, had not been easy. Aside from problems with the building of the telescopes and domes, there were other problems, and they started soon after the observatory complex began to expand.

When the road to the summit was first built, skiers began to take advantage of it to get to the snow at the top, but when work on the telescope began, the road was closed to the general public because of a worry about vandalism. The skiers felt deprived of a right they thought was theirs. Hunters soon joined the opposition as they became concerned about the disturbance that traffic up the mountain was causing to the sheep, goats, and wild pigs on the slopes below the observatory. The Hawaiian Audubon Society joined in the opposition and soon they had the support of the acting governor, George Ariyoshi. Mae Mull of the Audubon Society called for a halt to the building of further observatories. A "master plan" for the mountain was called for. This master plan was to be prepared by the University of Hawaii, and it was to look into all intended uses of the mountain, not just the scientific ones. The division of Forestry, the Fish and Game department, and the State Parks division were all to have their say.

When the first draft of the master plan was finally unfolded, Mull was not satisfied. She felt it wasn't going far enough in protecting native Hawaiian plants, and a rare and endangered species of bird on the mountain called the palila bird. The home of this bird is the mamane trees on the slopes of Mauna Kea. It is a small honey-creeper that feeds on the seeds of the mamane tree. Recent surveys indicate that about 600 of these birds nest in the trees on the slopes; they are not known to exist anywhere else in the world. Mull and her group were sure the observatories would affect the mamane trees, and therefore the palila bird. It was soon pointed out, however, that the real impact was coming from the sheep and goats on the mountain. They were eating the young mamane trees. A program was therefore set up to eradicate them; 20,000 to 30,000 goats and sheep were shot from helicopters, and as might be expected, this angered the hunters.

Mull called for a moratorium on observatory building, a "shakedown period" as she called it, during which no more observatories could be built. "Stop at six," she said. "Turn the spoilers away. The astronomers have had more than their share of the mountain." At this time there were six observatories on the mountain, three of which were still being built.

But there were many, including the Island Chamber of Commerce, who wanted more development, which would bring money to the island and help its fragile economy. As Jefferies and others battled these groups they stressed this. Jefferies also emphasized that astronomy was a "clean" industry, with no serious impact on the mountain.

Strangely, a large number of objections were directed at the appearance of the observatories on the top of the mountain. The observatories could easily be seen from Hilo on clear days. Some people referred to them as white "warts." Others said they got "bad vibes" when they looked up and saw them. A number of sportsmen suggested they should be built down in the bowls of the circular cones so they couldn't be seen. Jefferies pointed out that this wasn't feasible. Others suggested they should be put in silos and lowered out of sight during the day. Jefferies said that this would be extremely expensive and would create heat problems that would destroy the seeing. Still others argued that the domes should be painted blue to blend in with the sky. White has always been considered the best color for a dome because dark structures tend to collect more heat during the day. Jefferies offered to experiment with light blue on one of the smaller domes.

There were, however, large numbers of people who supported development of the observatories, and once the master plan was approved, opposition faded for a while. It quickly resurfaced, however, in the early 1980s when an expansion of the facilities at the midlevel site, Hale Pohaku, was planned. As the observatory grew it became increasingly apparent that a much better midlevel facility would be needed to house astronomers during their observation runs. Because the observatories are at such a high altitude it is essential that the observers do not return to a low altitude during their observing runs. As mentioned earlier, it has been found that they can easily acclimatize to the thin air at the top if, after observing all night, they spend their days sleeping at Hale Pohaku.

Several of the above groups opposed the construction of a new and larger facility at Hale Pohaku, and old wounds were soon reopened. One of the arguments centered around the large number of mamane trees that were in the area where the expansion was planned. The opposition was eventually overcome, however, and the facility was built. It was dedicated in 1983, and has since been named for the Hawaiian astronaut, Ellison Onizuka, who perished in the Challenger accident in 1986.

THE JAMES CLERK MAXWELL TELESCOPE

In the late 1960s astronomers in Great Britain began looking into the possibility of building a large telescope that would be capable of detecting radiation in the region between infrared and radio waves. It was a region of the electromagnetic spectrum that had not been investigated because of the specialized and costly instrumentation needed to detect it. Radiation in this region is given off by cool clouds of gas and dust. Since stars are born in clouds of gas, such a telescope would give us considerable information about the birth of stars. Gaseous nebulae and galaxies also radiate in this region.

A telescope designed for these wavelengths would resemble a radio telescope more than an optical telescope. Radio telescopes generally have much larger primaries, which are composed of mesh wire or metal, rather than glass. In the telescope the British envisioned, the primary mirror would be composed of panels of steel.

In 1975 a formal proposal was submitted for a 15-meter telescope. It was to cost approximately $9 million. In 1977 the project was given high priority and a committee was formed to look for a site. Mauna Kea was high on the list because its altitude allowed access to the region between infrared and radio waves. This region is now referred to as the submillimeter region. There was considerable opposition at the time to further development of Mauna Kea so the committee looked at a site in La Palma, Spain. It was finally selected as the best site, but there were problems signing an agreement with Spanish authorities.

For a period of about three years the project was at a standstill, then in early 1980 a review panel was set up to take a second look at it. They decided to cut $2 million from the budget. This was a serious setback to British astronomers, but a year later the Netherlands decided to join the project. They would supply 20% of the required funds in return for a proportionate amount of time on the telescope, so the project was back on line.

A new committee was set up to look at the sites again. By this time the opposition in Hawaii had subsided and they were invited to locate next to the UKIRT. An agreement was signed with the University of Hawaii in 1982.

Work on the dome began in the spring of 1983, and by September the base was complete. The dome itself was constructed in England. It was shipped to Hawaii in early 1984 aboard a Danish ship, the Odin Ace. It left England on February 14 and was expected to be in Hawaii about a month later. The ship did not, however, go directly to Hawaii, as expected; it went to Rotterdam and took on some additional cargo—20 tons of dynamite.

Delays continued, as people waited in Hilo for the arrival of the dome. Then the Danish shipping company claimed that they hadn't been paid, and wouldn't deliver until they were. To make matters worse, Hilo authorities found out the ship was carrying dynamite and wouldn't allow it to dock with the dynamite on

Dome of the James Clerk Maxwell telescope.

board. The dynamite was finally taken off in Costa Rica, but there was still the problem of the unpaid bill, and when the ship finally arrived in Hilo harbor it wouldn't dock until the bill was paid. Furthermore, they said there would be an additional charge of $3000 a day while they sat offshore. Fortunately, the problems were cleared up and the dome was off-loaded on May 12.

The dome was reassembled on the mountain during the summer of 1984. The telescope reached Hawaii during the summer of 1985, and by early 1986 it had been installed and was ready. First light came in December of that year. In early 1987 JCMT officials were approached by Canadian astronomers, offering to buy a 25

Dish of the James Clerk Maxwell telescope.

percent share of the telescope in return for an equal amount of observing time. At the insistence of the Canadians a fund was set up to restore a number of features that had been pruned from the project. An agreement was signed with them on April 26, 1987. The next day the telescope was formally dedicated by Prince Phillip, the Duke of Edinburgh. It was named the James Clerk Maxwell telescope after the early physicist who was born in Edinburgh in 1831. Maxwell made many contributions to physics, but is best known for his formulation of the laws of electromagnetism.

The headquarters of the JCMT was located with the UKIRT headquarters in Hilo. It is now referred to as the Joint Astronomy Centre.

THE CALTECH SUBMILLIMETER TELESCOPE

Although the JCMT is the largest submillimeter telescope on Mauna Kea, it is not the only one. Almost a year before it was dedicated, a smaller 10-meter telescope was completed by Caltech. Its shiny dome, looking like something out of the 21st century, sits only a few hundred yards away from the JCMT on what is known as Submillimeter Ridge.

CSO, as the observatory is now known, was the brainchild of Caltech physicist Robert Leighton. Leighton's interest in astronomy began in the 1960s when he built one of the first infrared telescopes. In the late 1970s he turned his attention to the submillimeter region. He realized it would take an extremely accurate surface to detect these waves, but it was a challenge he couldn't pass up.

With grants from the National Science Foundation and the Kresge Foundation, he went to work. He began by making a backup structure of lightweight hollow steel tubing, and attaching 84 hexagonal aluminum honeycomb panels to it. Each of these panels was independent of the other. He then moved the axis of the dish to a vertical position, rotated it, and machined it to the approximate required shape—a mathematical curve called a para-

Dish of the Caltech submillimeter telescope. View from the back.

The Caltech submillimeter telescope dome.

bola. Waves reflected from such a curve all focus to the same point.

He then epoxied aluminum sheet metal to the surface of this and machined and polished it. And finally, to allow for fine adjustment, he placed three adjusting screws behind each panel. "Bob and I spent hours in the early days crawling around on the structure adjusting these screws," said Walt Steiger, who was site manager from 1987 to 1992. "With three screws on each of 84 panels, there were a lot of adjustments to make."

The detectors, or receivers, which must be kept at liquid helium temperature, were designed by Tom Phillips. Phillips worked for ten years at Bell Labs where he specialized in milli-meter and submillimeter detectors. He came to Caltech in 1980, shortly after Leighton began working on the dish; he is now the director of CSO.

Everything was assembled and tested in Pasadena before it was brought to Mauna Kea—even the dome. After testing it thoroughly they took it apart and shipped it to Hawaii where it was hauled up the mountain piece by piece and reassembled. "The weather on the summit is much colder and windier than in Pasadena, so there were a few problems getting everything to work perfectly at first," said Steiger. "The huge shutter has been one of the major headaches. It creaks and groans a lot, and at times you hold your breath wondering if it will stay together." He chuckled. "But it's still running." Then smiling, he added, "Leighton and Phillips did a wonderful job. It's a marvelous instrument."

Strangely, although he made many trips to Hawaii and worked diligently to get the telescope working properly, Leighton never used it for research. He was content to build it.

Walter Schaal and David Vail also had a hand in the building. Schaal designed the hydraulic system that rotates the dome and dish. Vail machined many of the components. Interestingly, the building housing the telescope rotates with the telescope, but is totally detached from it.

The observatory was dedicated on November 22, 1986. The headquarters was located at the community college in Hilo, but when the college expanded, it was moved to a new site near the Hilo Mall.

Including two smaller telescopes with 24-inch mirrors built by the University of Hawaii while they were building their 88-inch reflector, the mountain now had eight telescopes on it. But the real coup for the University of Hawaii was still to come.

CHAPTER 4

The Largest Optical Telescope in the World—Keck

The University of California was an early leader in astronomical research. With the completion of the 60-inch reflector at Mount Wilson, then the 100-inch reflector a few years later, however, it fell behind the California Institute of Technology (Caltech). And when the 200-inch reflector at Palomar was inaugurated in 1948 it was completely overshadowed.

The University of California regained some of its earlier prestige in 1959 when the 3-meter (120-inch) reflector at Lick Observatory on Mount Hamilton was completed. For a few years it was the second largest telescope in the world. Then in the 1970s a number of telescopes were built, including one in Russia that surpassed the Palomar telescope (in size, but not quality—it has had many problems), and by the late 1970s the 3-meter reflector at Lick Observatory had slipped to tenth in size. In addition, the once dark skies over Mount Hamilton were deteriorating because of light pollution from nearby cities. Lick tried to stay at the forefront by devising better detectors, but it was soon evident that something had to be done. A faculty meeting was therefore called in 1977 to consider the matter.

For several years plans had been floating around for a 3-meter telescope to be located in the California coastal mountains, but nothing had come of them. The possibility of resurrecting this proposal was discussed at the meeting. Sandra Faber of Lick Observatory was present at the discussion. "We were all electrified when Joe Wampler suggested that we were fooling around with a size of telescope that wasn't exciting," she said. "He thought it might be easier to get a lot more money for a much larger telescope—one about 10-meters in diameter. I was excited about the proposal right from the beginning."

Soon after that meeting, a meeting of the Santa Cruz and Berkeley astronomy departments was called to consider the proposal. Faber remembered bringing some calculations to the meeting showing what could be achieved with a 10-meter telescope. At the meeting a committee was set up to look into the proposal further.

EARLY HISTORY

One of the people asked to serve on the University of California committee was Jerry Nelson of Lawrence Berkeley Laboratory. The committee discussed the various alternatives, and decided that a larger telescope was in their best interest. Nelson was asked to look into the problems that might be encountered. Was it possible to build a telescope as large as, say, 10 meters? And, if so, how should they proceed?

Born in Glendale, California, in 1944, Nelson went to school in southern California. His interest in astronomy developed early, but he was also good at mathematics, and in high school he decided to become a mathematician.

Upon completion of high school he went to Caltech with the intention of becoming a mathematician, but during his first year he took a physics course from Nobel laureate Richard Feynman. Feynman taught the course only once, and from it he developed and published his famous *Feynman Lectures*—three large red volumes well-known to all physicists. "It was a marvelous class. I loved every minute of it," said Nelson. He loved it so much, in fact, that he decided to switch to physics.

Despite his decision to major in physics, astronomy was still

Jerry Nelson.

close to his heart, and when he got a chance to work on an astronomy project he took advantage of it. Robert Leighton and Gerry Neugebauer were building one of the first infrared telescopes, and over the next couple of years Nelson worked with them on the project. This was his first taste of telescope making, and he thoroughly enjoyed it. He machined parts, worked on the mirror, helped build the dome for it on Mount Wilson, and when it was finished he spent two summers making observations. Yet, strangely, he didn't think seriously about astronomy as a career. "Gerry Neugebauer advised me not to go into astronomy," said Nelson. "He told me there was a better future in particle physics."

Nelson graduated from Caltech in 1965 and went to the University of California at Berkeley to work on a Ph.D. in particle physics. "I learned a lot in particle physics, and I don't have any regrets about going into it," he said. "It was really better training for what I'm doing now than astronomy. Particle physics has a lot of engineering in it. I had to build a lot of equipment, learn electronics, and learn how to debug. They were useful skills for me later on."

Nelson graduated with his Ph.D. in 1972, and went to work for Lawrence Berkeley Laboratory. At the time he was asked to serve on the telescope committee in 1977, he was working in astrophysics.

The first thing he did after being asked to look into a large telescope was learn everything he could about telescope making. He had some experience with the 62-inch infrared telescope but still felt like a novice. He was soon absorbed in the technology of telescope construction. There were obviously two routes to a large telescope. The mirror could be monolithic (single) or segmented (made up of many small mirrors that act as a single mirror). Nelson looked into the pros and cons of both designs. The segmented mirror concept seemed to have many advantages. Small mirror blanks were readily available, the equipment for polishing and aluminizing them was relatively cheap compared with that needed for a larger mirror, and small segments would reduce the weight of the mirror. Furthermore, a damaged or destroyed segment could easily be replaced whereas a damaged monolithic mirror would be disastrous.

The casting and annealing of a large, thick mirror blank was out of the question. It had taken almost a year to cool the mirror blank for the 200-inch telescope. A monolithic mirror would have to be thin, but even if it was thin, it would still sag under its weight, and would have to be supported. The worst thing, however, would be the transportation of a large mirror to its site.

Nelson soon became convinced that a segmented mirror was the only way to go. He also realized that he would need help in looking into the details of constructing a large segmented mirror.

He therefore recruited another physicist at Lawrence Berkeley Laboratory, Terry Mast. "I was talking to Terry almost every day about the problems of the telescope," said Nelson. "He became so interested I finally asked him to join me."

Like Nelson, Mast had also gone to Caltech for his undergraduate degree. He was one year ahead of Nelson but saw him frequently. They lived in the same dormitory, and also in the same "alley," as they referred to the dormitory hallways.

"We were a year apart, and at Caltech you do little more than study, so we didn't interact very much," said Mast. When they graduated, however, they both went to the University of California at Berkeley for their Ph.D's. Furthermore, they both went into

Terry Mast.

particle physics, and although they weren't in the same group, they got to know one another quite well.

After they graduated they both stayed on at Lawrence Berkeley Laboratory. They continued working in particle physics, but Nelson eventually began to have second thoughts about particle physics. "The particle physics groups kept getting bigger and bigger, until finally I asked myself, 'Do I really want to do this?' Things seemed so complex and involved, and there were so many names on papers," said Nelson. He began to look around, and soon his early interest in astronomy began to resurface. He found himself intrigued with small dense stellar objects such as pulsars, neutron stars, and black holes, and he gradually drifted into astrophysics. One of the first projects he became involved with was an x-ray binary (double system) with a 1.7-day period called Hercules X-1. Both x-ray and optical pulsations had been detected coming from it. Nelson and his group looked at the optical variations of the system, and discovered that the period of the variations wasn't constant. They also noticed that the period of the optical variations was different from that of the x-ray variations. After considerable work Nelson and his group were able to disentangle the data and come up with a model of what was going on. Using this model they were able to show that the mass of the x-ray source was approximately 1.3 solar masses. It was a significant result, and one Nelson was justifiably proud of.

About this time Mast also began turning his attention to astrophysics. He began working part-time on a project related to pulsars, then later got involved in a project on cosmic rays.

Nelson continued his work in astrophysics as he began working on the design of a telescope, but soon found he had little time for astrophysics. Mast followed the same route. "Both of us changed our groups," said Mast. He went on to explain how this was possible. "One of the really nice aspects of Lawrence Berkeley Lab was that there was a tremendous variety of research there, and it was easy to get involved in several different things. If you got interested in something some other group was doing, you just walked down the hall and started attending their meetings. There was a lot of fluency, and many people switched groups." He paused. "I think it was that kind of environment that allowed Jerry to do what he did. If he had been at a strictly academic institution, he would have been much too involved in teaching, committees, and academic affairs to do it. And if he had been in a commercial or industrial lab he would have worked on a project that was assigned to him."

Nelson and Mast were soon joined by George Gabor. Gabor brought an extensive knowledge of hardware and electronics to the group. The three men worked long hours, and most weekends in the early years. They were spurred on, to some degree, by a competition. The committee decided that both the segmented mirror and monolithic mirror concepts should be investigated. Nelson's group was to submit a proposal for a 10-meter segmented mirror, and a group from Santa Cruz was to submit a proposal for a monolithic mirror.

"We were young and cocky," said Mast, reminiscing about the era. "We were sure that our proposal would be better." But the group from Santa Cruz, which included Joe Wampler and Dave Rank, had considerable experience, and had built some fairly sophisticated equipment, so there was no guarantee they would win. "We were lucky, though," said Mast, shaking his head and smiling. "I don't think they had the environment and facilities that we had. We had tremendous access to engineers, technicians, and machinists. You could ask them for advice, and even ask them to build things for you. Many of them got quite excited about the project."

I asked Nelson how he happened to come up with the segmented mirror concept. He laughed. "It wasn't a new idea," he said. "There are no really new ideas. I often tell people that the guy who invented bathroom tiles is the guy who invented the segmented mirror." Radio astronomers had, indeed, used the concept for years. Their radio wave "collectors" are made up of many small "dishes." In radio astronomy it is relatively easy to do this because radio waves are much longer than visible waves. Radio waves from several separate dishes can easily be brought together and superimposed. With visible waves, which are a million times shorter, it isn't easy.

Nelson and his group soon realized that one of the most difficult problems would be aligning the mirrors so they would act as a single mirror.

STRESS-MIRROR POLISHING

One of the early decisions was to make the mirror of Zerodur, a glass/ceramic product made at Schott glassworks of Mainz, Germany. Zerodur is a composite of materials of negative and positive indices of expansion; when the materials are brought together the combination has almost zero expansion with temperature change.

One of the first problems Nelson and his group had to face was the shape of the mirror surface. If it had been spherical the problem would have been easily overcome. When a spherical surface is cut up in segments, they are all alike. If you cut up a spherical salad bowl, for example, each segment is the same as each of the others. It is therefore easy to make a mirror with a spherical surface: just polish the appropriate number of identical pieces and bring them together.

But a mirror with a concave spherical surface doesn't focus light to a point. Only a parabola—like the surface behind the lightbulb in your car headlights—does this. If you cut a parabolic surface into small segments and look at it carefully you see that each piece is different; furthermore, each piece is asymmetric. Grinding and polishing a hollow spherical curve into a piece of glass is relatively simple, but grinding a section of a parabola into it is extremely difficult.

Nelson saw the difficulties and decided to try a different approach. Many years earlier, telescope maker Bernard Schmidt had devised a technique for polishing nonspherical surfaces. All telescopes at that time were either refractors, which employ only lenses, or reflectors, which use a large mirror. The primary of a refractor is a large lens that "collects" light from a star; the resulting image is magnified by a smaller eyepiece lens. In the same way, the primary of a reflector is a large mirror. It collects light, which is reflected by a small secondary mirror to an eyepiece where it is magnified.

Schmidt wanted to build a telescope that combined lenses and mirrors. He determined that such a telescope would have a number of advantages over conventional reflectors and refractors. Schmidt soon found, though, that the refracting lens, or "corrector" in his design, needed a toroidal (like the surface of a tire) ripple in its surface. No one had ever polished a lens with such a surface before.

Schmidt considered the possibilities. He soon realized that if he could distort the glass, polish it, then release it, he might be able to get the surface he needed. He therefore took a thin glass blank and placed it over a pan which was smaller than the blank. After sealing it to the pan, he pumped the air out of the pan and the center of the glass was sucked into it, leaving a slightly concave surface. He polished the surface flat, and released the vacuum. The glass sprung back giving him the toroidal shape he needed.

With this lens Schmidt made the first of what are now called Schmidt telescopes. Many major observatories now operate large Schmidt telescopes, which are particularly valued for their wide field of view. They can also be made very compact, and are therefore a favorite telescope of amateur astronomers. Strangely, when Schmidt tried to market his new design, no one was interested, and he didn't live to see its popularity later on.

Nelson wondered if he would be able to use a similar technique on his mirror segments. Could he strain them, polish a spherical surface into the strained blank, then release it and get the required asymmetric surface? He talked to Nobel laureate Luiz Alvarez about his idea, and Alvarez directed him to Jacob Lubliner, a civil engineer. Brilliant and easy-going, Lubliner did most of his serious work in a cafe on the edge of the campus. "If you wanted Cobie [his nickname] you didn't bother to call him at his office," said Mast. "He wouldn't be there. You'd go across the campus to the cafe where you would find him talking to his students or working on some problem with a cup of coffee in front of him."

Lubliner was different from most civil engineers in that he liked to get a "feel" for a problem before he dug into it. According to Mast, he preferred to start with a "back of the envelope" estimate. He would go over all aspects of the problem in his mind until he thoroughly understood it. Only then would he go to the computer.

Nelson discussed the problem with Lubliner, and together they figured out exactly how each segment of the mirror would have to be stressed so that when a spherical surface was polished into it, it would relax to the proper section of a parabola. It was a difficult problem, and the two men worked for months on it. They were assisted later by Robert Weitzmann. "Cobie was a gifted guy . . . good with numbers," said Nelson. "It was really his mathematical expertise that led to the elegant stress-mirror polishing technique we used." The equations that the men had to solve were complex and difficult. They had to work out the appropriate equations for each section, then determine what forces to apply around the edge of the disk to get the deformation they needed.

They determined that the best way to apply the technique was to build a circular holder, or "jig," for the mirror with levers around it that could be loaded with weights. When weights were hung around it asymmetrically, and a spherical surface was polished into the blank, it would acquire an asymmetric surface when released. What made the work particularly difficult was the accuracy required. Each of the segments had to focus light to the same point to one part in 100,000.

A considerable amount of experimentation was required before the technique was perfected. By now it had been determined that there would be 36 segments in the mirror, and each of these segments would have to be polished in this way. But the final segments were to be almost 2 meters across, much too large to experiment with, so all of the early work was done on scaled-down segments, roughly one-quarter the size of the final segments.

Once the techniques were perfected they had to be taught to the commercial firms that would produce the mirrors. ITEK Optical Systems of Lexington, Massachusetts, and Tinsley Laboratories of Richmond, California, were selected.

PASSIVE AND ACTIVE SYSTEMS

Figuring and polishing the mirrors was not the only problem Nelson and his group had to surmount. When all of the segments were brought together, they had to act like a single mirror to an exceedingly high accuracy. So, while Nelson was working with Lubliner on the polishing of the mirrors, he was also working with Mast and Gabor on the problem of holding the segments together accurately. The major problem was gravitational sag. As the telescope was moved from position to position the surface of the mirrors would change slightly. Temperature variations would also create small changes.

The problem, therefore, was twofold. First, the mirror segments had to be attached to the telescope frame so that they would hold their shape and restrict optical deformations to acceptable levels. This is now referred to as the passive system. In addition, however, a computer-operated "active" system would be needed to keep the mirrors aligned relative to one another so that they acted like a single mirror.

The passive system provides axial (perpendicular to the mirror surface) and radial (along the direction of the surface) support. The axial support system consists of "whiffletrees" (a common term of the 19th century that refers to the pivoting crosspiece that allowed oxen or horses to move somewhat independently when pulling a wagon)—systems of levers or pivots that move horizontally, but not vertically—under each segment. Each of the whiffletrees has 12 short metal rods about the diameter of a toothpick that support the mirror. They are threaded into a hole on the back of the mirror. Each segment therefore has 36 axial supports.

Radial support, which stops the mirror from moving side-

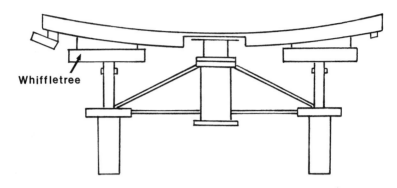

Cross section of the support system for a mirror segment showing the whiffletree.

Another view of the mirror subcell assembly. (Courtesy Keck Observatory)

ways, is provided by a central post. A cavity is drilled in the back of the mirror and a thin (0.011 inch) stainless steel disk attached to the top of the post is fitted into the hole and bonded to the glass.

When the mirror is pointed to the zenith (directly upward) the weight of the mirror is supported by the whiffletrees. When it is pointed toward the horizon its weight is supported by the thin disks. For positions in between, the weight is distributed between the two systems.

The active control system, an ingenious system devised by Nelson, Mast, and Gabor, is a major innovation in telescope design. It consists of two basic components: sensors and actuators. The sensors determine where the mirrors are relative to one another, and if they are out of alignment they send a message to a system of actuators to bring them back in alignment. The sensors, which are mounted two to a side on the back side of each mirror, consist of a flat "paddle" on one mirror which fits between two parallel plates on the adjacent mirror. There is a 4-millimeter gap on each side of the paddle. A thin plate of gold is bonded along the top and bottom of each gap, and a charge is introduced onto the plates so that the gap acts as a capacitor. If one mirror moves relative to the other, the paddle moves up or down in the gap. This increases the size of one of the gaps, and decreases the other, which in turn decreases or increases the electrical capacitance of the capacitor. When a change in capacitance occurs a message is sent to a computer that calculates how much of an adjustment is needed to bring the two mirrors back in line, and the actuators bring them back. The capacitance at each of the sensors is measured twice a second. With two sensors along each of the hexagonal sides (except for the outer edges) there are a total of 168 sensors on the mirror.

The actuators, 108 in all, are attached to the base of the whiffletrees. Each actuator moves the mirror by turning a finely machined screw. As the screw turns, a nut moves along a threaded shaft and presses against a hydraulic bellows. The screw can be turned as little as 1/10,000 of a rotation. This moves the nut about 4 millionths of an inch. Incredible as it may seem, this is too much

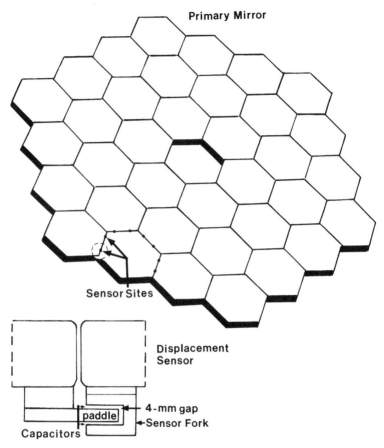

The segmented primary mirror. The displacement sensor is shown at the lower left.

for accurate positioning. To increase the precision, a hydraulic lever was added that reduced the change by a factor of 24. With this, changes of less than 1 millionth of an inch can be made.

Mast and Gabor both worked on the active system. Mast was concerned with converting the capacitance to a number that could be fed to the actuators that moved the mirrors. Gabor did much of the work on the sensor system. There was, however, input into all

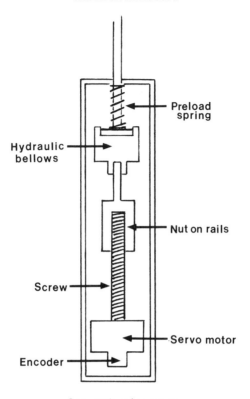

Cross section of an actuator.

aspects of the systems from each of the men; it was a joint effort. "We looked at a lot of different ways of doing these things," said Nelson. "We made many models, tested them, rejected many, accepted others. There was a lot of trial and error."

THE BUILDING OF A PROTOTYPE

Until 1979 there were still two proposals: one for a monolithic mirror and one for a segmented mirror. Proposals from each group were presented to the committee, and after thorough study they decided to go with the segmented mirror. "That was a real milestone in the project," said Nelson. "Before that we were just working on a proposal."

The technology had been demonstrated on a quarter scale model. Now it was time to move to a full-scale model. The first step, however, was a prototype. In it a full-sized segment would be used along with a small piece representing an adjacent segment. This project began in 1981. One of the people assisting was Barbara Schaefer. Schaefer received her bachelors degree in physics and astronomy from the University of Wisconsin. After graduating she went to Kitt Peak as a telescope operator. She stayed there for four years, then went to Hawaii to work as a telescope operator on the IRTF. In 1982 she got a chance to join Nelson's group and she took it; they had just started building the prototype when she arrived.

A full-sized hexagonal segment was used in the prototype. Three actuators were placed under it, along with four displacement sensors along one edge. They were placed between the main segment and a small reference mirror. A testing system was set up that produced interference fringes—closely spaced straight lines—across the disk and its adjacent reference mirror. If the lines of the segment continued straight across the reference mirror, the two were properly aligned.

In the first tests the mirror had a spherical surface; in other words, it wasn't part of a parabola. At this stage the group was concerned only with whether the active system would work.

Schaefer referred to the building of the prototype as one of the high points of her association with the telescope. "We took interferograms of the hexagonal segment and its little neighboring piece and saw the lines going straight across the edge," she said. "Someone would walk up and push on the main mirror and the lines would distort. Then the control system—the sensors and actuators—would quickly bring it back in line, and the lines would straighten out. It was fun to watch and it gave us a real sense of accomplishment."

Not everything went smoothly, however. The group got a scare when they tested the first completed mirror. The mirror blank comes from the manufacturer in the form of a circle. It is stress-polished, holes are drilled in the back for the supports, then the edges are cut off so it is in the form of a hexagon. They hoped that all of this would not change the figure that had been polished into the surface. But, alas, it did. The mirror segment relaxed slightly—only a tiny amount—but it was significant enough that it had to be corrected for. Would they have to repolish it? After looking into it, they decided a better approach was to give the mirror a slight permanent push, enough to distort it back to its original shape. This was done with "warping harnesses." They are a set of small aluminum leaf springs that are adjustable to $1/25,000$ of an inch that lie near the top of the mirror-positioning actuators.

As they got more experience in mirror polishing they were able to predict and to some degree compensate for the small changes. Later, another technique called ion figuring was used to help reduce it. Ion figuring is a technique that was developed by a group at the University of New Mexico. The exact shape of the surface required is fed into a computer. Information from the computer goes to an "ion milling machine"—a device that knocks off high spots by spraying them with ions. This device is mounted on a stage that can be swept across the surface of the mirror. "Ion figuring has made a big difference," said Peter Wizinowich, the optics manager at Keck. All of the ion figuring is now done for them by Kodak. Even with ion figuring, however, the warping harnesses are still needed for final fine-tuning.

LOOKING FOR FUNDING

While the research on the mirror and telescope was going on the question of how it would be funded hung over them. They had to find a donor. But first they needed an estimate of how much it would cost to produce.

Gerald Smith of JPL had guided the NASA infrared telescope

through to completion, and was now project manager for the IRAS satellite program. A University of California Executive Management Committee responsible for guiding the development of the telescope approached him asking for his advice. Would it be possible to make some cost estimates and set up an organization plan? Smith was heavily involved with IRAS but he told them he would find time to help them. Over a period of about three years he set up an overall plan for the building of the mirror, telescope, and dome, which included budgeting and scheduling.

When IRAS was launched in 1983 Smith was available and the Executive Management Committee asked him to work full-time as project manager. The project was still unfunded at this point. The University of California had been trying for about three years to find money, but had been unsuccessful. Finally in 1984 they got a promise of partial funding from the Hoffman Foundation. Approximately $50 million was available to them. Smith's estimates showed that it was about $20 million short, so the University of California asked Caltech if they were interested in coming in as a 20 percent partner.

Caltech was, indeed, interested, and it looked as if the project would go. Furthermore, the inquiry to Caltech paid a double dividend. The Keck Foundation had funded several Caltech projects in the past and, through Caltech, they got interested in the telescope. And to the delight of everyone involved, they decided to fund the entire project (or at least $70 million of it). The money, however, would come through Caltech, so they would no longer be a junior partner; they would be an equal partner.

Negotiations began between the University of California and Caltech, and a nonprofit corporation, the California Association for Research in Astronomy (CARA), was formed. Caltech and the University of California would have equal representation on the board. According to the agreement, Caltech would pay for the telescope through the Keck grant, and the University of California would pay for developmental technology and the operating costs for the first 25 years. The agreement was signed in 1984. The Keck grant of $70 million was the largest single gift ever given for a scientific project.

At the time of the agreement they had both the Hoffman grant and the Keck grant. Since the total from the two grants was considerably more than they needed for a single telescope they began talking about a second telescope—a twin to the first. They could be operated as a unit through a technique called interferometry. Soon, however, there were problems with the Hoffman grant, and the money was returned.

Once funding was available and everyone knew the telescope was going to be built, things went into high gear. Smith began putting together a project team. He set up offices in the basement of the physics building at Caltech, but was soon cramped and had to move to offices adjacent to the Caltech campus on Wilson Street. An engineering and administrative staff was put together to design, build parts of the telescope, and issue contracts; soon over 20 people were working in the offices. The years 1984 to 1989 were busy ones for Smith and his group. The telescope and all of the key hardware, along with the dome and building had to be designed, and key contracts had to be issued. ITEK Optical Systems and Tinsley Laboratories were given contracts for polishing and fabricating the mirror. TIW Systems of Madrid, Spain, was given the contract for fabricating the telescope structure, and Coast Steel of Vancouver, British Columbia, was given the contract for the dome.

DESIGN OF THE TELESCOPE

Although the mirror is the most important part of a telescope, it must be mounted in a rigid structure that properly supports the optics and scientific intruments, and is able to point the mirror in any direction with high precision and accuracy. The design of the telescope and mount were therefore of supreme importance.

The mount finally chosen is called an altitude-azimuthal mounting. It is a simple but reliable design. Most of the design work was done by Jacob Lubliner, who earlier worked on the mirror, and Stefan Medwadowski. (Interestingly, Medwadowski and Lubliner were from the same town in Poland. They left and went their separate ways for many years, but both ended up working on the telescope—even on the same part of the telescope.)

There are several parts to the telescope. First, there is the mirror support, or mirror cell. It has a complex steel lattice structure, as shown in the figure. It is necessary that it be strong, as it has to support 14 tons of glass, 6 tons of segment support, and 4 tons of instruments. It was designed by Medwadowski. Second,

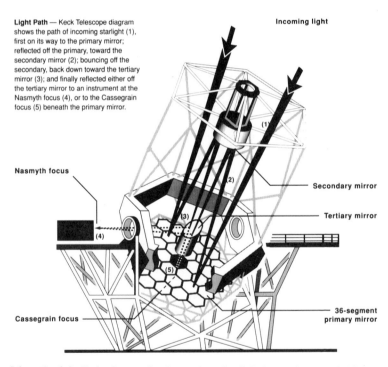

Light Path — Keck Telescope diagram shows the path of incoming starlight (1), first on its way to the primary mirror; reflected off the primary, toward the secondary mirror (2); bouncing off the secondary, back down toward the tertiary mirror (3); and finally reflected either off the tertiary mirror to an instrument at the Nasmyth focus (4), or to the Cassegrain focus (5) beneath the primary mirror.

Incoming light

Nasmyth focus

Secondary mirror

Tertiary mirror

(4)

(5)

36-segment primary mirror

Cassegrain focus

Schematic of the Keck telescope showing an incoming light beam. (Courtesy California Association for Research in Astronomy)

there's the tube of the telescope, which also has a lattice structure. One of its main features is to hold the secondary mirror, but it must also hold several tons of instruments, and keep the optical pieces of the telescope accurately collimated. Light from the primary mirror is focused back to the secondary mirror, which in turn reflects it through the hole in the center of the mirror, or to a tertiary mirror which reflects it to instruments on the Nasmyth platform.

The telescope tube rests on hydraulic bearings that define the elevation axis. They allow the telescope to be moved up and down. These bearings are attached to the top of the yoke, which carries the telescope load to the pier. In addition to withstanding this weight, the yoke has to be stiff enough and strong enough to withstand pressure from strong winds on the telescope. Two platforms, called Nasmyth platforms, are also attached to the yoke. Each of these platforms must be capable of holding 10 tons of instruments.

The bearing of the azimuthal axis on the bottom supports 159 tons on a thin layer of oil. The telescope is driven in the azimuthal direction by four relatively small ½-horsepower motors. All motions of the telescope are computer controlled.

Most of the mechanical parts—the drives, gears, auxilliary pier supports and so on—were designed and built under the direction of Hans Boesgaard. He came to the project in 1984 as a consultant, and after CARA was formed he became structural manager of the Keck telescope. Boesgaard had earlier worked on the University of Hawaii's 88-inch telescope, and the NASA infrared telescope. Born and raised in Copenhagen, Denmark, he came to Montreal in 1953, then after short stays at DuPont in New York and Kaiser Engineering in California he took a job as instrumentation engineer at Lick Observatory. He has been working on telescopes ever since.

By the mid-1980s work was well under way on the Keck telescope. There was still the problem, however, of where it would be located.

CHAPTER 5

The Continuing Story of Keck

In the early 1980s a committee was set up to select a site for the 10-meter telescope. Chaired by Robert Kraft of Lick Observatory, it consisted of people from the University of California campuses at Berkeley, Los Angeles, Santa Cruz, and San Diego. They were to look for a site that had a minimum of cloud cover throughout the year, low background light, excellent seeing, low water vapor, and reasonable access. Funds were limited so they weren't able to test each of the sites in detail, but they gathered as much information about them as possible. Thirteen sites were considered. Several were rejected quickly, others less quickly, until finally the list was wittled down to five: La Palma in the Canary islands, a mountain near Madeira, Spain, Junipero Serra Peak in California, Mount Bancroft, also in California, and finally, Mauna Kea in Hawaii. White Mountain in California was also tested in some detail.

Early on, the most seriously considered peak was Junipero Serra. It had good seeing, but was only 5500 feet high so it gave little access to the infrared. Four Indian tribes claimed the mountain as their territory, so months, and finally years, were spent in courts trying to resolve the problems. They could not come to an agreement and the University of California finally gave up. Another peak in California, White Mountain, also looked good. It had good seeing one year, but the next year it had deteriorated considerably, and they decided not to take a chance on it.

Prior to 1979 there were little available data on Mauna Kea. The University of Hawaii's 88-inch telescope was operating, but for the first few years there were problems with it. In 1979, however, three telescopes went into operation and within a short time there was considerable information about the seeing, and conditions on the summit. It soon became clear that Mauna Kea was superior to any of the other sites, and it was selected.

BUILDING THE DOME AND TELESCOPE

Work on the design of the dome began in August 1985; the groundbreaking for its foundation came in September. Smith remembered it as one of the highlights of his association with the telescope. He had been with the project for two years at this point, and was still stationed in California. It would be several years yet before he would move permanently to Hawaii. "It was an exciting event," he said. "There were a lot of officials here from state government, and from California. Several of them gave speeches." After so many years of planning, construction was finally beginning. It was a big moment for Smith and the other members of the project team.

The dome had to be designed to allow for the harsh weather conditions on the summit. It had to withstand winds of up to 145 miles per hour and allow operation of the telescope in winds up to 65 miles per hour. Furthermore, it had to be insulated so that the temperature inside was as close to the ambient temperature outside as possible.

At 100 feet high and 122 feet across, it is slightly smaller than the dome of the 200-inch telescope at Palomar, even though it houses a telescope twice as big. The reason is the short focal length of the Keck mirror; a large dome is not needed.

Weighing in at 700 tons, the dome arrived in Hawaii in May 1987. Flatbed trucks took it to the summit piece by piece where it was assembled. Work on the building adjacent to it had begun before the telescope arrived, and it was now almost complete. It

Building the dome for Keck I. (Courtesy Keck Observatory)

Constructing the base for the Keck I dome. (Courtesy Keck Observatory)

Steel girders going in place on the Keck I dome. (Courtesy Keck Observatory)

would house the control room, computer facilities, mirror aluminizing facilities, electronics shop, and instrument assembly shops.

About the time the dome was arriving in Hawaii, fabrication of the telescope was beginning at TIW Systems in Spain. To make sure it was up to specifications, it was partially assembled and tested before it left Spain, then it was put on a boat for Hawaii. It came into Kawaihae Bay on the Kona Coast, just a few miles from Waimea, the town in the north part of the island that had been selected as headquarters of the Keck project.

"The arrival of the telescope in August of 1989 occurred about the same time our staff moved here," said Smith. "It was an exciting moment when the ship pulled into the harbor with the telescope structure aboard. It was partially assembled so we could see large sections of it." And just as the pieces of the dome had been hauled up two years earlier, the partially assembled sections of the telescope were trucked up. Many of the personnel from Pasadena were now coming to Hawaii. The headquarters buildings had been started about a year earlier and were now almost finished. They were therefore ready when most of the scientists, engineers, and others arrived.

Many people participated in the assembly and testing of the telescope. One of them was Bill Irace of Jet Propulsion Laboratory in California. As chief systems engineer he was responsible for the organization and coordination of the project. He also spent a considerable amount of time helping align the primary mirrors. Irace had spent several years working on space telescopes at Jet Propulsion Lab before coming to the Keck project.

"There were several aspects to the construction . . . structural, optical, electronics, and software. Hans Boesgaard was responsible for the structure, Mark Sirota, the electronics, and Hilton Lewis, the software. My main job was to coordinate them," said Irace.

Ron Laub from the University of California was also hired soon after production began. As facility manager he would supervise the technicians working at the summit. Laub had been indirectly associated with the telescope for several years; he was on the review committee for the design of the observatory and the headquarters building. Born and raised in California, Laub spent much of his youth on a farm. "I tried farming for a while when I was young," he said. He admitted, though, that it was a short-lived endeavor. "It was too difficult to compete with the big boys, so I gave up and went to college." His first stop was Cal State in Fresno. Later he attended Foothill College in Mountain View and City College in San Jose, where he majored in engineering.

His first job was with Varian Associates in Palo Alto. While at Varian he learned optical coating techniques for aluminizing mirrors. It was a skill that would be invaluable to him later. From Varian he went to Lick Observatory. "I thought I would only be there a year or so," said Laub. Then with a laugh he added, "I ended up staying 22 years. It was a much more challenging job than I thought it would be, and after a few years I was superintendent of the site." He still has a large picture of Lick Observatory on his wall.

Ron Laub.

When Laub arrived in Hawaii the dome was complete, but they were still in the early stages of constructing the telescope. At this point there was no commercial power at the summit, and this became his first job. He was eventually responsible for stock, transportation, scheduling, hiring, and many other things. "I was up at the summit about three days a week at the beginning," he said. Lately, with paperwork and so on, he has been spending more and more time at the headquarters in Waimea.

The first mirror segment arrived in Hawaii in January 1990.

Because of his experience with optical coating, Laub was also soon aluminizing the segments. I asked him how he went about it. "They have to be cleaned first," he said, "and that takes a lot of elbow grease. You have to be very particular because the surface has to be extremely clean. You then put it in a vacuum system and pump it down. Originally that took about an hour and a half; we can now do it in about an hour." He sat back and relaxed, then continued. "We use aluminum for the coating. Silver has some advantages, but it oxidizes too fast, and you would have to recoat it frequently. However, the Institute for Astronomy has discovered that silver survives about two years and as a result it is being considered for Keck II."

FIRST LIGHT

In October 1990, the first mirrors were installed on the telescope. "When we had the first mirrors in place we started to check the system to see if we could get them all pointing right, so they would act as a single mirror," said Smith. "With three mirrors in place we got an image." The addition of each new mirror seemed like a milestone to the group. Finally they had nine mirrors in place, which was the equivalent of the Hale telescope, and they took their first picture. The image, which was obtained using a CCD (charge-coupled device) camera, was of the spiral galaxy known as NGC 1232 in the constellation Eridanus.

As they continued putting mirrors in place, they used what is called the Hartmann test to check their alignment. In this test Hartmann screens—foam pads punched with 19 holes—are used to cover each mirror. With the screens in place, the telescope is aimed at a star. Light from the star passes through the holes and is reflected back out. A camera, placed just in front of the focus of the mirror, picks up the rays and records where they strike. If they don't strike exactly where they are supposed to, the mirrors are adjusted until they do.

The adjustments are incredibly small—as tiny as 1/25,000 of

The mirror cell with 9 of the 36 mirror segments in place. The first photograph was taken at this time. (Courtesy California Association for Research in Astronomy)

an inch. They are made to the warping harness under each of the mirrors.

In April 1992, the 36th mirror finally went into place. "That was the high point for many of us," said Smith with a broad smile. "Everyone was on the mountain." Except for testing and adjusting, the telescope was now complete.

While work on the mirror was continuing, technicians were rushing to finish other parts of the telescope. In addition to a primary mirror, a telescope also needs a secondary mirror to reflect the light to the observer. Two different secondaries will be used with the Keck telescope, depending on the type of observations being done. At optical wavelengths an f/15 (the f number gives the ratio of the size of the mirror to its focal length) secondary will be used, and for infrared studies an f/25 secondary will be used. The Keck telescope also uses a tertiary mirror in the center of the primary.

The f/15 mirror was polished by opticians at the University of California at Santa Cruz. It is 57 inches in diameter, weighs 1250 pounds, and, like the primary, is made of Zerodur. It arrived in Hawaii in July 1992. When the telescope is used for infrared studies the f/15 secondary will be replaced by the f/25 secondary. Both are held in a steel frame approximately 50 feet above the telescope's primary mirror.

Light from the secondary can be passed through the hole in the center of the primary, or it can be reflected to what is called the Nasmyth deck by use of the tertiary mirror, as shown in the figure on page 86.

Most of the optical problems are now the responsibility of Peter Wizinowich, the optics manager, who came to Keck in October 1991. Born in Winnipeg, Canada, Wizinowich got his bachelor's and master's degrees at the University of Toronto. It was there that he also got his first taste of astronomy. "The university has a telescope in Chile. It's only a small telescope—24 inches in diameter—but it's well-instrumented. So it's great for graduate students. You could get a lot of observing done on it. I was resident astronomer down there for about a year." When Wizinowich completed his master's degree he went to the Canada–France–Hawaii telescope for four years as an optical technician. He soon realized, though, that if he wanted to advance, he needed a Ph.D.; he therefore went to the University of Arizona where he worked under Roger Angel, well-known for his innovative new techniques in mirror making. Wizinowich's thesis was on stress-lap polishing and optical testing.

"I wasn't here for first light at the prime focus," said Wizi-

Peter Wizinowich.

nowich, "but I was here when first light was obtained using the f/15 and f/25 secondaries." He grinned, then pointing to his right eye he said, "First light from the f/15 hit this eye."

Wizinowich and Nelson have been concerned over the past year or so with optimizing the optics, in other words, with getting the best resolution possible. Resolution is a measure of how well you can distinguish two objects. It is measured in seconds of an arc (the circle is divided into 360 degrees, with each degree consisting of 60 minutes and each minute of 60 seconds). Most observatories have a hard time getting below about 1 second arc seeing; half second arc seeing, however, is common on Mauna Kea. This is equivalent to distinguishing the two headlights of a car at 500 miles.

"We've had resolutions down to 0.7 arc second," said Nelson. "Whether that was the seeing or our optics we're still not sure. A couple of months ago we had a night of phenomenal seeing. Some of the individual segments indicated resolutions of 1/4 arc second."

Wizinowich also talked about that night. "Getting images like that when the telescope is nowhere near optimized was exciting," he said. I asked him what the telescope would eventually be capable of. "It's limited by the atmosphere," he said. "We want to make sure it's not limited by the optics. We want it so that when the seeing offers itself we get 1/4 arc second seeing."

DEDICATION

The dedication of the telescope took place on November 7, 1991. One hundred and fifty people from Caltech, the University of California, and the Keck Foundation were invited. They were joined by the Keck staff and a number of others.

It was a bright, clear windy day as everyone streamed into the dome in jackets and sweaters. A few wispy clouds sailed high overhead. The 40 degree temperature was moderate for the mountain, but chilly for anyone without a coat. Exactly one half (18) of

The Keck telescope seen with the dome open. (Courtesy Keck Observatory and Roger Ressmeyer)

The Keck telescope (Keck I). (Courtesy California Association for Research in Astronomy and Roger Ressmeyer)

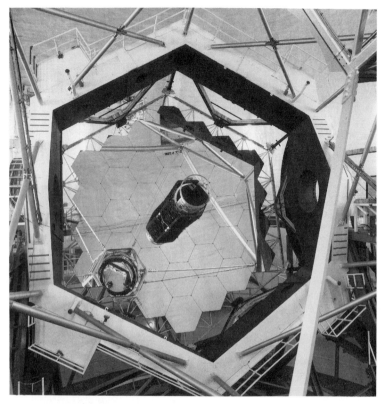

The Keck mirror with all 36 segments in place. (Courtesy California Association for Research in Astronomy)

the 36 hexagonal mirror segments had been installed. The visitors sat in folding chairs on the floor beneath the giant telescope; the observatory staff stood along the catwalk overhead.

The speeches were made from a dais next to the telescope's pier. The ceremony began with an invocation—a Hawaiian chant given by Kalena Silva of the University of Hawaii. He blessed the telescope, and everyone who had worked on it. The telescope sat motionless in a horizontal position as several speeches were made; with the completion of the ceremony it came to life; 150 tons of glass and steel began to rotate soundlessly over their heads. Then the dome began to turn. A number of people looked down at the floor, wondering if they were moving. When the telescope got to the shutter it began to rise upward. The crowd broke into an applause as the shutter started to open and sunlight streamed in.

When the Keck money was first given to Caltech and the University of California, the money from the Hoffman grant was still available, and plans had been made for a second 10-meter telescope to be placed beside the first one. The two were to be tied together via interferometry, so they could be operated as a unit. But when the Hoffman grant was returned, plans for the second telescope were shelved. In April 1991, however, the Keck Foundation announced it would fund the second telescope. So, when Keck I, as the original 10-meter telescope is now called, was dedicated, plans were already being made for Keck II. In fact the traditional Hawaiian ground blessing for Keck II took place immediately after Keck I was dedicated. The crowd went from the dome to a roped-off area at the far end of the observatory building, where Keck II would be built. Several of the dignitaries lined up with o'o sticks—early Hawaiian digging implements. A Hawaiian chant was conducted while they tried to look like they were digging up the red cinders.

The dedication ceremony was followed by a dinner the next day. It was held at the Mauna Kea Beach Hotel a few miles from Waimea on the ocean. The emcee was Walter Cronkite, formerly of *CBS News*. "We were pleased that Walter came," said Smith. "He also attended the ceremony on the summit the day before the dinner, and he didn't appear to have any problems with the altitude."

The keynote speaker at the dinner was Ed Stone, Caltech vice president and professor of physics, director of JPL, and chairman of CARA. Stone talked about the important role the telescope would be playing in the future of astronomy. He pointed out several problems that would be attacked using it, placing partic-

The courtyard at Keck headquarters with a section of lawn showing the size of the Keck mirror.

ular emphasis on the question of the origin of the universe, and how solar systems form.

William Fraser, senior vice president for academic affairs at the University of California and vice chairman of CARA, talked about the funding of the telescope by Keck, and the founding of CARA. And keeping with tradition, Cronkite signed off with, "And that's the way it was."

THE INSTRUMENTS OF KECK

A telescope is only as good as the instruments that are attached to it. Besides CCDs that give an image of the object, the major type of instrument that will be used with Keck is the spectrograph. A spectrograph spreads out the light from a star into its colors, or wavelengths, and gives a series of dark lines. These lines occur because atoms and molecules in the star's outer atmosphere absorb light of certain wavelengths. The spectrum of a star gives us information about the makeup and the physical properties of the star, and is therefore invaluable to the astronomer.

Keck will have two spectrographs. Both are state of the art, and were designed using complicated computer analysis. The first is a low-resolution imaging spectrometer. "It was designed for high sensitivity," said Smith. "So we'll be using it to look at very faint, distant objects. This is one of the things the telescope was designed to do. It will really be at the frontier as far as distant galaxies and quasars are concerned." In addition to obtaining spectra, the low-resolution imaging spectrograph will be able to produce detailed images of objects.

Steve Vogt of Lick Observatory and his group are working on the most expensive of the instruments planned for the telescope: a $3.1 million high-resolution spectrograph. With Keck's tremendous light-gathering power, this spectrograph will produce the most detailed spectra ever obtained from faint objects such as quasars.

These are the two major optical instruments of the telescope. They will allow astronomers to obtain spectra in minutes that would take other telescopes hours. Keck's remaining three instruments are directed at the infrared. Because of its accessibility to the infrared, and its size, the Keck telescope will quickly become the world leader in infrared astronomy. Two of the infrared instruments are cameras: a near-infrared camera for short wavelengths and a far-infrared camera for longer wavelengths. Both will be thousands of times more sensitive than infrared instruments produced only 20 years ago. The infrared instruments used at Palomar during the 1980s, for example, employed a single solid-state detector. Keck's near-infrared camera will contain 65,000 similar, but much more efficient detectors.

The final instrument is an infrared spectrograph. Other instruments such as photometers (that measure light) and CCDs will also be used.

"Each of the five instruments is designed specially for this telescope," said Smith.

KECK II AND INTERFEROMETRY

At a press conference at Caltech in April 1991, Howard Keck, the chief executive of the Keck Foundation, announced that they would fund a second 10-meter telescope—a twin to Keck I. "The technological breakthrough achieved with the construction of the first Keck telescope has encouraged us to push further the boundaries of astronomical research that will only be possible with a second telescope," said Keck. The Keck Foundation offered to pay approximately $75 million toward the construction of Keck II. This will cover about 80 percent of the projected cost of the telescope.

Keck mentioned at the news conference that he had no real interest in astronomy, and was only interested in making worthwhile scientific grants. I asked Smith about this comment. He was silent for a few moments, then said, "I think what he meant by that is that he's not an amateur astronomer and isn't very knowledgeable about astronomy. He did appear to be very interested in the purpose of the telescope, and its technological development."

Keck II will be located about 285 feet from Keck I. They will be linked together by a technique called interferometry, and as such will have a resolution, but not the light-gathering power, of a 285-foot mirror. The linking of telescopes via interferometry has been around for many years. Radio telescopes have been linked since the 1950s. Interferometry is much easier with radio waves than with visible waves, however, because radio waves have a much greater wavelength.

In interferometry, signals from two or more sources are brought together and superimposed to give a strengthened signal. It is critical, however, that they are brought together so that the wavelengths are exactly in phase. This occurs naturally only when the telescopes are pointed directly upward, and the signals are brought together at a point midway between them. In most cases the signal from one arm of the interferometer will take longer to reach the midpoint than the signal from the other. This means that one signal has to be delayed slightly. The technique has been used successfully in the near infrared at Mount Wilson. Two small mirrors, 31 meters apart, were used in this project. Many other similar projects are now in progress; one is at Lowell Observatory and another at Georgia Tech and University of Georgia.

The details of the Keck interferometer have not been worked out yet. "There's a lot of technology that needs to be developed," said Smith. "The long optical path—285 feet—between the two mirrors is long even for the infrared. We will have to have very precise delay lines if we are to bring the two signals together properly."

The two Keck telescopes will be supported in a few years by four smaller "outrigger" telescopes. They will have a mirror diameter of 1.5 to 2 meters, and will improve the imaging capability considerably. According to present plans the four smaller telescopes will be capable of being moved among any of 18 fixed stations. Signals from the outriggers, Keck I, and Keck II will be routed to a beam-combining room in the basement of Keck II.

Steel girders for the Keck II dome being taken off a barge in the harbor at Hilo.

Building the Keck II dome.

Keck II is scheduled to be completed in 1996 and interferometry between them should follow within a couple of years. The
outriggers, however, are not likely to be completed before the year
2000.

Current plans call for interferometry only in the infrared.
How far away is interferometry at optical wavelengths? Smith
said he didn't see much hope of making it work in the near future;
he felt that it would be an extremely difficult challenge. Ken Wei-
ler of the Naval Research Lab, who has been working in the area,
said that optical interferometry is like "trying to read a newspaper
through a bubbling aquarium while jumping on a waterbed."

It may be a few years off, but it will no doubt come.

ADAPTIVE OPTICS

Interferometry, when perfected for the infrared and visible,
will give astronomers a tremendous advantage in seeing. But it is
not the only technique that will help. Another technique called
adaptive optics is now at the forefront of astronomy. The original
idea came from Horace Babcock, who is now at the Observatories
of the Carnegie Institution of Washington. Babcock realized that
much of the restriction on seeing for any telescope is related to our
atmosphere. As the light from a star makes its way down through
the atmosphere it is continually shifted. It acts as if it has come
through many small lenses, each of which focuses it to a different
point; this causes the beam to "dance" and results in a blurry
image. Because of this, the very best resolution that can be
achieved is about 1/2 arc second, which is about ten times the
theoretical limit for a "best" image. In other words, it should be
possible to reach resolutions of 0.05 arc second. In 1953 Babcock
began looking for a way to attain this value.

In the case of a perfect image the "wave front" is a straight
line (it is actually an arc of a very large circle). Turbulence in the
atmosphere, however, distorts this line. Astronomers hope that

adaptive optics will allow them to straighten it out. In adaptive optics, mirrors are used that detect the tilts and bumps in the light beam's wave front and compensate for them. A slight tilt, for example, can be corrected by use of a flat mirror tilted in the opposite direction; a bump can be smoothed out by using a thin flexible mirror that has actuators beneath it to change its shape. With the use of such a system it is possible to bring the blurred image back into focus.

Crucial to the system is a reference star. If the actuators in the adaptive optics system can keep the "test" star sharp, the rest of the stars in the field will automatically be sharp. Unfortunately, the reference star has to be fairly bright, and all fields of view do not contain a bright star. Because of this, artificial stars are used. Most artificial stars are now made using lasers. Laird Thompson of the University of Illinois developed a system in which a pulse of ultraviolet laser light was fired into the sky in the direction the telescope was pointed. The atmosphere reflected the beam and created an artificial star. Unfortunately, reflection did not occur at the top of the atmosphere—only about 12 miles up—and the back-scattered rays of light were not perfectly parallel. More recently, however, higher-powered lasers have been used and they can create artificial stars out much farther.

A simple adaptive optics system called HRCam has been used with the Canada–France–Hawaii telescope for several years. It makes rather crude corrections, but they have significantly improved the seeing.

Adaptive optics will also be used with the Keck telescopes, but the details have not been worked out yet. "We're just at the point of coming up with a strategy and a management plan for adaptive optics," said Wizinowich. "It's a hot topic right now and a lot of people have opinions on how it should be done." Although a committee has been set up at Keck, no firm decisions have been made.

Simple adaptive optics devices are already in the system. Tilt corrections, for example, can be made. The infrared secondary mirror has a very fast response, so it is possible to sense wave-front tilt errors and correct for them now.

I asked Wizinowich how good he expected the seeing to be once the adaptive optics system was in place. He said they hoped to get it down to 0.05 arc second, and possibly as low as 0.02. This is equivalent to distinguishing the two lights of a car 16,000 miles away.

PROJECTS

Keck I will be ready for its first projects in mid-1993. As the world's most powerful instrument at both optical and infrared wavelengths, astronomers will be able to use it to look back almost to the beginning of the universe. They expect to see quasars 12 billion light-years away, which is 3 billion years after creation. Furthermore, when Keck II comes in line with Keck I in 1996 astronomers expect to see galaxies or quasars as they were 1 billion years after creation. Together, the two telescopes will have eight times the light-gathering power of the 200-inch reflector on Mount Palomar.

What are some of the questions astronomers are hoping to answer with the telescope? One of particular importance is: How did the universe get its present large-scale structure? On a very large scale, the universe is mottled—strings of superclusters appear to be strung around huge spherical voids in space. The Keck telescope may help us understand how and why this came about.

Another important question is: What is the universe made of? Much of the universe is known to be composed of dark matter, but astronomers still do not know what form this dark matter takes. With Keck's tremendous capacity for obtaining spectra of faint objects, we may find an answer.

Plans are also under way to search for extrasolar planets. It is extremely difficult, if not impossible, to see planets directly. Because of this, astronomers generally look for a "wobble" of a star caused by a planet. Several stars that exhibit wobbles have been

One of the first pictures taken with Keck I. (Courtesy California Association for Research in Astronomy)

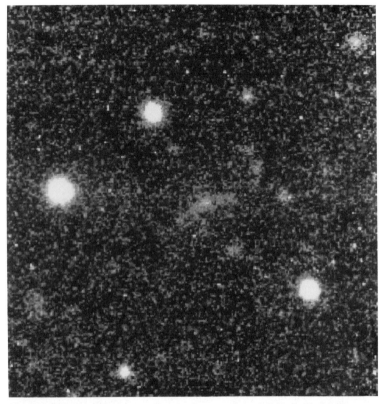

The object in the center is the most distant galaxy ever imaged. Taken by Keck I. [Courtesy Keith Matthews (Caltech) and California Association for Research in Astronomy]

found. Using Keck, a search will be made of the nearest 100 stars to see if any have Jupiter-like objects orbiting them.

Astronomers will also search for clues to star formation. How do stars form? How do planetary systems form? How did our galaxy, the Milky Way, form? What is at the core of our galaxy? And other galaxies? Many exciting discoveries no doubt await us in the first years after Keck goes into operation.

CHAPTER 6

Visiting the Top of the World

At just under 14,000 feet, Mauna Kea is the highest point in the Pacific. It's also higher than most of the peaks in the Rocky Mountains. Looking at it from a distance, you might not appreciate its height; it's not jagged like many of the peaks of the Rockies. In fact, with its rounded top it looks like an oversized molehill. But looks are deceptive; it's a high mountain. Furthermore, when you drive up it, you're starting at sea level. In the Rockies you're usually starting well above sea level.

FIRST TRIP

After watching Mauna Kea from Hilo for several days, seeing it covered with snow, then slowly begin to melt off, I was eager to visit its summit. I had read much about the hardships of some of the early climbers—altitude sickness, loss of memory, and difficulty thinking—and I naturally wondered how the altitude would affect me. But most of all I was anxious to see the tremendous array of telescopes on the summit.

I made arrangements through the Joint Astronomy Centre in Hilo for the trip. Kevin Krisciunas was to be my guide. A marathon runner with thick wavy hair, Krisciunas phoned the morning we were to leave to tell me there would be a delay. The four-wheel vehicle we were to have used had been in an accident. The previous driver, who was from England, had been driving down the middle of Saddle Road, the road leading to the observatory, when a car suddenly appeared over the crest of the hill in front of him. Forgetting he was in America, he swerved to the left.

It was raining heavily as we left Hilo. I was in the back seat; Kevin sat in the front with an astronomer from England who had insisted on driving. I was a little apprehensive, but he assured me he had done considerable driving in the United States, and wouldn't make the same mistake the previous driver had. I was also assured by Kevin that it wouldn't be raining at the summit. The summit does, indeed, get some precipitation, but it is mostly in the form of snow in the winter. During December and January it usually gets a couple of blizzards, which can leave a few feet of snow, and a storm had, indeed, passed over the mountain about a week earlier.

The final few miles to the summit are steep and treacherous in places. Astronomy faculty therefore use four-wheel drives for the trip. There is, in fact, a sign at the midlevel facility saying that only four-wheel drives are allowed beyond it. Furthermore, all rental agencies specify in their contract that their cars are not to be taken on Saddle Road.

Rain continued to batter the windshield as we made our way through the streets of Hilo toward the entrance to Saddle Road. Kevin kept us entertained by challenging us with some of his scientific puzzles. The traffic, as always in downtown Hilo, was heavy. Not being thoroughly familiar with the city, I wasn't sure where we were until I saw the sign "to Saddle Road." I knew then that we weren't far away. Flowers, ferns, and trees lined the road, with a house here and there. We passed a sign that said "Kaumana Cave." I filed it away in my mind as a place that might be worth visiting later.

Finally we reached Saddle Road. I had heard so much about it. Was it as treacherous as they said? It had been built during World War II to give access to a military installation in the center of the island. As its name suggests, it is a road in the "saddle" between the two major volcanoes of the island; it is, in fact, the only major road across the interior of the island. As we entered, a sign read, "Turn on your lights."

The first few miles were smoothly paved, but you soon felt as if you were on a twisting roller coaster. There was no such thing as a straight stretch—it was one hill after another, and one curve after another. To my surprise there was considerable traffic on the road, and I could see that it would be relatively easy to have an accident. Still, it wasn't nearly as bad as I had anticipated.

The rain finally began to ease up; it was now more like drizzle, but the air was still thick and humid. Kevin was doing most of the talking, telling us about the observatory and some of the people who worked there. I watched the scenery around me; I had come prepared to take pictures but there didn't seem to be much to take pictures of. The vegetation was dense and wet, and like most regions on this side of the island, it looked like an impenetrable jungle.

After about ten miles the blacktop stopped; only the center section remained smoothly paved. When the road was originally built, it consisted only of this center section. After the war, however, when the public began using the road, a work crew painted a white line down the center of this section and patched the ground on either side of it with asphalt. They obviously didn't prepare the ground properly, however, as the patched sections were now full of potholes and cracks. Most people therefore prefer to drive on the smooth center section—as long as no cars are coming. The problem is that there are so many twists and turns, you never know when cars are coming.

I tried to take a few notes as we went along so I wouldn't forget anything, but the vehicle bounced and jerked so much it was impossible. As we continued to climb, the scenery around us changed dramatically. From sea level to about 6000 feet there was dense vegetation. Parts of this region get as much as 300 inches of rain a year; Hilo itself averages about 150, so the entire region is green and fertile.

Gradually, however, the vegetation began to give way, and the landscape was covered with low trees and shrubs. The most common trees are koa and ohia. There was considerable game here at one time, namely sheep, goats, and wild boar; except for a few strays now, however, only the wild boar remain, the sheep and goats having been shot off.

Suddenly we came upon a huge lava field that extended for miles on either side of the road. I wondered where it came from. Mauna Kea is, of course, a volcano, but for all practical purposes it is extinct, its last eruption coming 3000 years ago. On the left side of Saddle Road, however, was Mauna Loa, and it is still active; lava flowed down its sides as recently as 1984, but the lava around us wasn't from 1984. It was, however, from Mauna Loa, as I found out later, coming from an eruption that occurred in the mid-1800s.

Finally we passed out of the lava flow and back into a region of trees, shrubs, and grass. It was an open area, with groves of trees interspersed with grasslands. At the 28-mile mark we turned off Saddle Road onto a well-paved and relatively straight road. According to Kevin it was only a few miles from here to the midlevel facility at Hale Pohaku. Cattle were grazing on the dry, grassy slopes around us, and in the distance a few old structures were visible.

The road was now much smoother and wider, but within a short time it became evident that we were climbing at an even greater rate than we had been previously. Our driver shifted to second, then finally to first. As our vehicle ground its way slowly up the sweeping curves I studied the land around us. Ecologically, this is one of the most interesting areas of the mountain. Mamane and naio trees are abundant, and there are bird and animal species that are not found anywhere else on Earth. The palila bird that I mentioned earlier, is here. Hawks, geese, and game birds such as quail and partridge also populate the area, and there are wild boar.

It appeared barren and dried out, and indeed it does get little rain. The sky overhead today was clear, except for a few drifting clouds. Finally, a long structure with many small peaks appeared through the low brush and trees ahead. We were at the midlevel facility. It had recently been named for the Hawaiian astronaut Ellison S. Onizuka who perished in the Challenger accident.

HALE POHAKU

The view from the veranda of the dining hall at Hale Pohaku is breathtaking. Below is Saddle Valley, studded with stunted trees and grasslands, and beyond in the distance is Mauna Loa. Looking in the direction of Hilo you usually see the top of the clouds associated with the inversion layer, and indeed this was the case today. And as Kevin had promised, we had driven out of this layer and were now well above it.

It isn't always clear in this region, however. The section of road from Saddle Road up to Hale Pohaku can be treacherous because of the clouds that drift through the region. They sometimes envelop you in a dense fog. I drove down from Hale Pohaku several times in a fog so thick I could barely see the line on the road ahead of me.

The facility at Hale Pohaku is reminiscent of a moderate-sized alpine hotel. From the veranda of the dining room you can see the sleeping quarters of the astronomers below, with long wooden walkways leading to them. These rooms are needed because the high altitude of the summit doesn't allow astronomers to remain there for 24 hours. Furthermore, once they are acclimatized, they can't return to a low altitude during their run (which can last for several days). Hale Pohaku, at 9200 feet, is an ideal altitude for astronomers during their sleeping hours, and time off, since there is much more oxygen here than at the summit. Astronomers therefore spend their nights on the summit, come down to the midlevel facility to sleep and eat during the day, then return to the summit the following night without any difficulty.

For anyone going to the summit it is necessary to stay at Hale Pohaku for at least an hour. People sometimes skip this layover, but they usually end up regretting it. Without acclimatizing you

Sleeping rooms for astronomers at Hale Pohaku.

View of the sleeping quarters from the dining hall at Hale Pohaku.

View of the clouds associated with the inversion layer. A radio telescope (part of VLBA) is seen in the foreground.

A section of the road near the summit.

can feel miserable at the summit. The hour helps get you used to the thin air, and prepares you for the even thinner air at the top.

The surroundings at the midlevel facility are pleasant. Music streams into the dining room from the kitchen, and several of the walls are decorated with striking photographs of the domes at the summit. A balcony overlooking the dining area contains pool tables along one side, and a small library at the end. Sliding doors lead onto a deck that has several lounging chairs on it. It looks inviting from the inside, particularly in that it is sunny much of the time, but when you go out, you quickly find out it's not ideal for sun-tanning. The temperature at the midlevel facility, while higher than at the summit, is still cool. You need a jacket if you plan on spending much time outdoors.

The 72 rooms that house astronomers, technicians, and others are small and have tightly fitted blinds. Days and nights are turned around for these people, so the rooms are made as comfortable as possible. Despite this, most people take a day or so to adjust to the turnaround.

Above the dining room are the offices and computer rooms that are available to astronomers. An astronomer's day usually begins late in the afternoon. Most rise about midafternoon and do the required computer work for their night run; they eat between 4 and 6 and leave shortly thereafter for the summit.

APPROACHING THE SUMMIT

After an hour's stay at Hale Pohaku, we donned jackets and started our trip to the top. It is approximately 8 miles to the summit from here, but the road is much steeper than the road we'd just come up. Krisciunas was now our driver. We also had some new passengers, namely Dr. Joseph Tatarewicz of Maryland, and his wife and son.

We passed the sign saying that only four-wheel-drive vehicles were allowed beyond this point, and it was easy to see why: the road was steep and unpaved. As steep as it was, though, it was considerably less steep than the road Herring and other early observers had used. Krisciunas pointed out the original road to us as he drove. It was so steep it was hard to believe that anyone could have driven up it. (Most of the route to the top is the same as the original route; it differs only near the bottom.)

All vegetation ceased just above Hale Pohaku. Above 11,500 feet, there wasn't even a blade of grass. Still, the view below was spectacular. We could even see Haleakala on the neighboring island of Maui, though Hilo was still buried beneath the inversion layer. We were now in a different world; fields of lava, rocks, and gravel were on either side of us. It was strangely reminiscent of the surface of Mars, and in fact, some of the rocks and sand were red, as on Mars.

It appears harsh and barren, yet there is considerable life here; most of it, however, is small. The only large vegetation found above 11,500 feet is an occasional silversword, the amazing spike-leafed plant that blooms only once in its relatively long life. Mostly there are lichen and moss. The animals are mainly arthropods, namely spiders, mites, centipedes, moths, and other bugs.

As we continued up the mountain, patches of snow began to appear. The entire mountaintop had been covered with snow only a week earlier, and looking at it from below, I had anticipated that it was a couple of feet deep near the summit, but the snow we saw around us lay in patches; many regions had no snow at all. The sides of some of the cinder cones were completely covered, but with the exception of a few snowbanks, the snow appeared to be only a few inches deep. This is the only region in Hawaii where you can ski. For the mainland skier, however, it may be a bit of a disappointment; there are no lifts, so you have to walk back up the hill after you have skied down it. Actually, there are places where you can park your car, ski down a slope, and arrange for someone to pick you up and take you back to the top. Skiers from some of the other islands, particularly Oahu, scurry to the Big Island after a good snowfall. There were no skiers on the hill that day, although many tracks were visible on the hillsides.

The snow itself seemed strange and different to me. In many

Several of the domes as seen from the summit. Keck is on the left. The Canada–France–Hawaii dome is in the center, with the University of Hawaii and United Kingdom domes on the right. The Caltech dome is in the foreground.

Snow on the summit. The Keck dome is on the right.

places iciclelike shapes appeared to grow out of the ground. It was obvious that the main way the snow disappeared was through sublimation. No water was running down the hillsides, despite the 32°F temperature and bright sunlight. Some water does come from the melting snow, however. Near the summit is a small lake called Lake Waiau that was created from melting snow. You can't see it from the main road, but it is surprisingly large, about 3 acres.

Also near the top is the adz stone quarry. Early Hawaiians climbed here to get stone for some of their tools. They sharpened the rocks from this quarry by chipping them. Then they bound the sharpened stones to sticks to make axes, hoes, and other tools.

The last four miles of the road are now paved. The traffic up and down the road during the day is relatively heavy, and before the road was paved a cloud of dust was left after each vehicle— dust that interfered with seeing. The road was paved to keep the dust down. There is, unfortunately, a disadvantage to the smooth pavement during the winter: it is much slipperier than the old unpaved road.

ON THE SUMMIT

As we drove the last mile to the summit we saw steel signs flattened to the ground. They were the signature of the 100 mile an hour winds that had raked the area only a week earlier. Several of the signs that still stood had holes in them, drilled purposely to stop them from being blown over.

Reaching the summit we jumped out. The first thing I noticed was that it wasn't as cold as I thought it might be. I had a long-sleeved shirt and vest on, and thought they might do. Kevin cautioned me that I would need a jacket inside the domes, so I put one on.

How did I feel in the thin air? There is, after all, only about 60 percent the amount of oxygen here there is at sea level, and I had read a lot about how strange some people feel. Headaches and dizziness were supposed to be common. I didn't have any trouble breathing, but it was easy to tell that I was at a high altitude. If I stood still I didn't notice anything, but the moment I began to walk I knew I wasn't going to be able to overexert myself. I was, indeed, a little light-headed, but aside from that, I didn't feel too bad, and within a short time I felt fairly comfortable. I had been told that young people—those under 20—suffer the most. And in our case, this appeared to be true. The teenager who was with us took only a short walk before he was back resting in the car.

The altitude affects different people in different ways. Some people become forgetful, some feel more confident, others do strange things. There are, in fact, many stories about the strange things people have done. A well-known science popularizer, for example, came to the mountain to do a TV show. By the time the equipment was ready and the show was to go on, he had almost passed out and was forced to cancel his part of the show.

Another incident that was the talk of the mountain happened one Sunday. An astronomer heard a banging on the locked observatory door. When he opened it a woman was standing before him in the nude—despite the freezing temperature. She asked if she could have a tour of the observatory. He asked her where her clothes were, and when she didn't appear to know, he rustled up some coveralls for her from the dome, then proceeded with the tour.

There are also stories about some of the workers who constructed the dome fatiguing easily, being disoriented, and not being able to make parts fit.

APPROACHING FROM WAIMEA

Not everyone going to the summit starts at Hilo. Keck and Canada–France–Hawaii headquarters are both in Waimea, in the northern part of the island. Astronomers from here approach Saddle Road from the other end.

On the day I came up from that end, it was sunny, with occasional light drizzle. Andy Perala, Erik Hill, Jay Pasachoff, and

I left Waimea for the Keck Observatory about 8:00 A.M. Perala, a public relations specialist at Keck, was our guide and driver. Erik Hill is a staff photographer for the *Anchorage Daily News*, and Jay Pasachoff, an astronomer. Perala had worked with Hill earlier in Anchorage; in fact, they shared (along with several others) a Pulitzer prize for a series of articles in the *Anchorage Daily News* about alcoholism in Alaska. It was Hill's first trip to the observatory.

The first part of the trip is along the road to Kona. Huge fields of grass and cactus belonging to the largest ranch in the United States, the Parker Ranch, lie on both sides of us. Cattle grazed on the slopes. As we turned off Kona Road onto Saddle Road we could see the ocean in the distance behind us. Saddle Road from

The University of Hawaii 88-inch telescope dome.

Andy Perala in the control room of Keck I.

this end is not as twisty as it is from the Hilo end, but it is also not as well paved.

We had left the drizzle behind now, and the skies were clear. Perala and Hill sipped water from a bottle to prepare for the summit. Pasachoff bombarded Perala with questions about the telescope as I sat in the back seat talking to Hill. The four-wheel-drive vehicle bumped along over the potholes and cracks in the road. After about an hour we passed Pohakuloa Military Camp; a red flag was up indicating that military exercises were under way in the area. Off to the left several helicopters and a plane sat on an airfield runway. The helicopters reminded Andy of a story about a pilot who had landed his helicopter near the observatory on the summit, and was not able to get it off the ground. He radioed for help, and another helicopter came up to assist him, but after a few minutes they discovered they couldn't get the second helicopter off the ground either. Finally they realized that the problem was the air: it was too thin to take off. Both helicopters had to be trucked down the mountain.

Just beyond the military camp was Mauna Kea State Recreation Area; it is used mainly by climbers and hikers. Several rental cabins dotted the slope above the main facility. For remoteness and solitude you could hardly find a better spot on the island.

It is about ten miles farther (from Waimea) coming in from this direction, as compared with coming in from Hilo. As in the earlier trip from Hilo, we spent an hour at Hale Pohaku, then proceeded to the top. Most of our time was spent observing and photographing the giant Keck telescope.

SURVEY OF THE OBSERVATORIES

There are now a total of nine observatories on the summit, with work beginning on two more. The light-gathering power of the telescopes presently on the mountain is greater than any other observatory in the world. The only one that comes close is the European Southern Observatory in Chile, but it has access only to mostly southern skies.

One of the major advantages of the Hawaiian site, in fact, is its access to both the northern and southern hemispheres. A telescope situated at the equator would, in theory, have access to all stars in both hemispheres. Mauna Kea, at 20 degrees north, has only a small fraction of the sky blocked off from it.

Somewhat below the summit, in what is sometimes called Submillimeter Valley, is the James Clerk Maxwell telescope (JCMT). My first impression of its dome was that it looked like a giant white coffee can. It seemed much cooler inside the dome than it did outside, and I was glad I had my jacket on. Part of this was no doubt related to the fact we were now shielded from the sun. The domes of all telescopes, however, are kept at a constant temperature, close to the nighttime temperature, so that the mirror doesn't have to come to equilibrium when the shutters are opened and observing begins.

The JCMT is used in the submillimeter region of the electromagnetic spectrum, and because of this it looks more like a radio telescope than an optical telescope. Its dish is 15 meters in diameter, larger even than the Keck telescope mirror, but it is made of metal. Radiation collected by this dish is reflected to a secondary reflector, which in turn reflects it through a hole in the primary to what is called the Cassegrain focus. Several different instruments can be set up to receive the signal here.

Two types of detectors are used with this telescope: bolometers, which measure the energy received over a broad frequency range, and heterodyne receivers, which have high resolution, in the same way your home radio does. Measurements are made at liquid helium temperatures (-270 C) to minimize noise from the receiver. Furthermore, radiation measurements are made, not only of the object of interest and the sky behind it, but also of an empty section of the sky. The measurement of the empty sky is subtracted from the previous measurement. This is accomplished through the use of a "rocking" secondary. The secondary is shifted, or tilted slightly, for the empty sky measurement, then back to the object. This is referred to as chopping.

A short distance away from the JCMT is the Caltech submillimeter telescope (CSO). Its mirror is 10 meters in diameter, and like the Maxwell telescope, it also looks like a radio telescope, in that its dish is metal. The CSO dome is a 200-ton silver bubble that contains three stories. Within these stories are a control room,

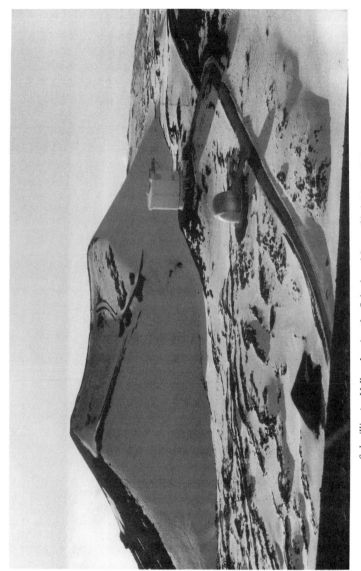

Submillimeter Valley showing the Caltech and James Clerk Maxwell domes.

electronics shop, a lounge and library, and a small galley. The telescope is controlled by computers in the control room; it is heated so that astronomers can work in relative comfort.

As in the case of the Maxwell telescope, its detectors are cooled to liquid helium temperature, and a chopping secondary is used to subtract the radiation from the background sky.

The facility is operated by a staff of ten based in Hilo. Headquarters are in Pasadena, where there is additional scientific and technical staff.

Looking upward from CSO at the surrounding ridge, one of the most prominent domes is that of the Canada–France–Hawaii telescope. It looks like a giant mushroom perched on one of the highest peaks in the area. The telescope itself stands on a concrete cylindrical pillar five stories high and 57 feet in diameter. The pillar is hollow and contains three rooms.

Looking at the telescope the first thing that strikes you is the size of the large horseshoe gear that serves as its north bearing. It is part of the mount that allows the telescope to track the stars. The telescope weighs 100 tons, the 142-inch mirror, 20 tons. It can be used in both the visible and the infrared, depending on how it is configured. It is equipped with three "head rings" that are interchangeable. A primary ring allows observations at the primary focus. In this configuration astronomers sometimes work in a small compartment at the top of the telescope, but most now prefer the comfort of the warm control room. The other two rings allow observations at the Cassegrain focus, just behind the central hole in the mirror. One is for infrared observations, the other for visible. A large crane inside the dome is used for changing the rings.

Once each year the mirror must be cleaned and recoated with aluminum. The operation begins with the detachment of the mirror from its cell. It is placed in a cart and moved out from beneath the telescope, then lowered with a crane to an aluminizing room below. This is normally a hair-raising event; one slip could result in a disaster of major proportions.

The Canada–France–Hawaii telescope dome. (Courtesy Canada–France–Hawaii Corporation)

Along the ridge from the Canada–France–Hawaii telescope is the University of Hawaii 88-inch reflector. With a hornlike projection on one side, its dome is the most distinct on the mountain. This projection contains a crane that is used for moving heavy equipment in and out of the dome. The telescope is mounted on a concrete pier that is isolated from the dome so that vibrations from the rotating dome will not be transmitted to the telescope. As in the case of the other telescopes, it is also controlled from a heated control room on the upper floor. Lower floors contain laboratory space, a machine shop and electrical shop, and a lounge and kitchen. There is also an aluminizing room.

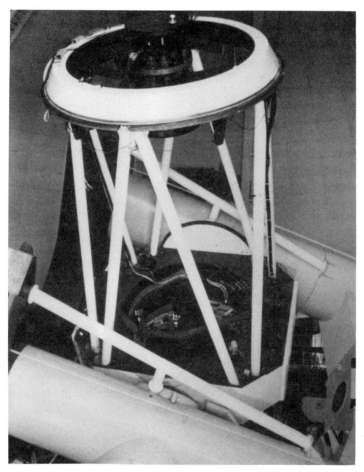

The Canada–France–Hawaii telescope. (Courtesy Canada–France–Hawaii Corporation)

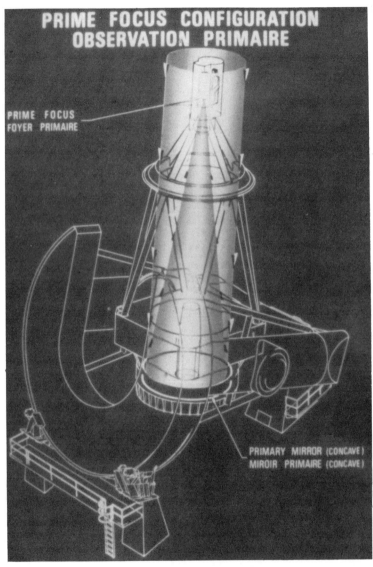

Schematic showing setup for prime focus observations. (Courtesy Canada–France–Hawaii Corporation)

Schematic showing setup for Cassegrain observations. (Courtesy Canada–France–Hawaii Corporation)

The University of Hawaii 88-inch telescope dome.

Many different instruments are used with this telescope; they can be placed at either the prime focus or the Cassegrain focus. A large Coudé spectrograph room has also been built to the south of the main building.

Also operated by the University of Hawaii is the NASA 3-meter infrared telescope (IRTF). It sits on a cinder cone a short distance away from the Canada–France–Hawaii telescope. It was built primarily for infrared observations of the planets. The most striking feature of this instrument is the massive steel yoke that holds the telescope. Each side is about 6 feet thick, with the overall yoke anchored to piers on the floor at both ends. This makes the telescope particularly rigid, so it can be set on stars very accurately.

Although most telescopes on the mountain can be used in both the visible and the infrared, this telescope is designed specifically for the infrared, and can therefore do things that other telescopes cannot.

When the observatory was first opened in 1979 only one point of the sky could be analyzed at a time. For a two-dimensional image the telescope had to be slowly scanned across the object, a process that could take hours. With more sophisticated and sensitive instrumentation, astronomers can now collect enough data for an image in about 15 seconds.

IRTF is only one-story, but includes a machine shop, electronics room, a lounge and control room; and the moving parts of the telescope weigh 137 tons.

The United Kingdom also operates an infrared telescope on the mountain. It has a diameter of 150 inches (3.8 meters). As we discussed earlier, it has an especially thin mirror that weighs only 6.5 tons. The telescope structure is also considerably lighter than telescopes of comparable size. And considering the fact that it contains a 150-inch mirror, its dome is a relatively small 60 feet in diameter.

Like all telescopes on the summit, the UKIRT dome and foundation are designed to withstand substantial earthquakes (earthquakes are common in Hawaii). An earthquake could, for ex-

The NASA infrared telescope dome.

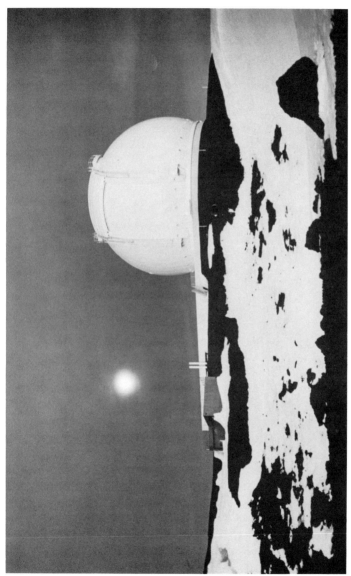

Dome of Keck I seen in the winter. (Courtesy California Association for Research in Astronomy)

ample, move the telescope laterally by several millimeters, but because of the way it is designed, it would take less than an hour to bring it back to its original position.

The open tube of the telescope is mounted in a rectangular yoke, a design similar to the IRTF, but not nearly as thick. The yoke itself rotates between two steel piers. Many different instruments can be used with the telescope, and as in the case of other infrared telescopes, the detectors are kept at liquid helium temperature, and the secondary is designed for chopping.

The giant of the mountain is, of course, the Keck telescope. Its modernistic-looking dome is 101 feet high and 120 feet in diameter. It is perched on a separate cinder cone, not far from the NASA infrared telescope. Work on Keck II is progressing and within a short time two identical domes will decorate the summit.

Finally, the University of Hawaii operates two small, 24-inch telescopes that are still extensively used. They were put into operation in 1968 and 1969 when the 88-inch telescope was being worked on. They can be used with several different kinds of instruments, including photometers, spectrographs, and infrared instruments.

The real thrill at the summit, however, is the night sky, and I now looked forward to my first visit at night.

CHAPTER 7

Night at the Observatory

There was a slight wind as we reached the summit. The sun was low on the horizon, and strangely shaped clouds billowed up from the generally even inversion layer below. Orange-red rays from the sun reflected from them, giving a spectacular view. The air was brisk and the humidity low, in sharp contrast to the region below us.

Bill Heacox of the University of Hawaii at Hilo and several of his students were planning to observe with one of the 24-inch telescopes. Bill fumbled through a large ring of keys looking for the key to the dome. He tried one and frowned. "I may have brought the wrong key ring," he said, as he tried a second one. The students, all bundled in parkas, hats, and heavy boots, looked skeptically at one another. Then a smile came over his face as one of the keys finally turned. He pushed the door open and turned on the light. The telescope was small compared with the others on the mountain, but it is still used occasionally. Bill had brought up a new CCD (charge-coupled device) that the university had just purchased. It would give a significant improvement over the cameras he had been using, and he was anxious to see how well it worked.

"Don't step on any of the tubes," Bill yelled as he began assembling the cooling system for the CCD. Several of his students worked on a small computer that was connected to the CCD, adjusting knobs and making various tests. Water for coffee was soon boiling on the small hot plate, and music from an old radio that was speckled with paint filled the dome.

Bill's breath steamed in the cold as he walked over to the eyepiece of the telescope and started hooking an adapter to it. Discovering that the CCD wasn't going to fit unless he made some adjustments to the adapter, he turned to the group, "I have to go up to the machine shop in the dome of the 88-inch." It was only a short distance up the hill, so I went with him. "My specialty is using odd parts to get telescopes working," he joked as we jumped out of the four-wheel-drive vehicle and headed for the door.

When we returned to the smaller dome everything was set up and ready to go. He only had to hook the CCD in. It was now getting dark; Venus was high in the western sky, shining as brightly as I had ever seen it, and almost overhead was the red planet Mars.

As the sky darkened its beauty transcended the mountain. Stars shone brightly and steadily; they were like diamonds of all sizes scattered randomly over a sheet of velvet. The constellation Orion was prominent, high overhead, with Sirius to one side.

Bill opened the shutter of the dome and moved the telescope so that it pointed out; then the adjustments, calibrations, and data taking began. As the hours wore on, my feet began to feel like ice, mostly because we were standing on concrete. Late in the evening I walked out of the dome and looked around. I could see the domes of the other telescopes dimly in the distance. They were dark, but inside astronomers were busy taking data. I wondered what each of them was doing. I knew none of them were looking through the eyepiece of a telescope; astronomers no longer look through telescopes; in fact, with a few exceptions, they no longer even photograph the sky. Furthermore, I knew they weren't shivering in the cold as we were. Astronomers now spend most of their time sitting in comfortably heated control rooms staring at TV screens and computer monitors. The major reason for this change is the CCD.

Stars and gaseous nebulae. (Courtesy Canada–France–Hawaii Corporation)

CCDs

With the introduction of CCDs in the 1970s and early 1980s, astronomy underwent a significant change. The breakthroughs that made this possible were made at Bell Labs in the 1960s. Scientists discovered that certain semiconductors such as silicon, indium antimonide, and gallium arsenide are responsive to light; in other words, when light strikes them, electrons are ejected. A photon of light hitting a silicon atom, for example, will knock one or more electrons from it. These electrons, when collected, produce a small current that can be amplified and measured. In short, a light signal can be transformed into an electrical current, and this current can be used to produce an image.

To see how this is done, it is best to begin with a photographic plate. When the photons from an object strike this plate they activate crystals in it, which, when developed, give an image of the object. We see an image because more photons strike some points than others, and the more intense the beam, the more intense the image at that point.

In the same way, a CCD chip consists of many little light detectors, or pixels, usually in a rectangular array. When this array is exposed to photons, they strike the pixels and create currents. The currents from all of the pixels are converted into electronic pulses that can represent zeros and ones. These numbers are then fed to a computer which, in turn, feeds them to a TV screen where the image can be reconstructed, and of particular importance, the more pixels you have, the clearer the picture.

CCDs have many advantages over photography. First, digital data can easily be manipulated by computer, and this allows us to enhance and manipulate the images. Second, CCDs are much more efficient than photographic plates. To get a photographic image, a photon must strike atoms in the plate, activating them, which is a very inefficient process. Only about 1 to 2 percent of the photons from an object actually activate atoms in the plate; the rest are lost. This is not the case with a CCD. Almost every photon that strikes a pixel is recorded, so they are close to

An infrared array. CCDs are similar in appearance. (Courtesy Joint Astronomy Centre)

100 percent efficient. That makes them 50 to 100 times as sensitive as photographic plates; much dimmer objects can therefore be detected using them.

CCDs are also much faster than photographs. With a CCD it takes only seconds or minutes to get an image that might take hours with a photographic plate. And finally solid-state arrays can be made that are sensitive to infrared and submillimeter radiation. There are, however, some disadvantages, but they are gradually being overcome. CCDs can't, for example, cover the wide area a photographic plate can. Each pixel is like a little light "bucket" that collects light, and the more light buckets you have the greater the amount of light you can collect. To approximate a photographic plate you would need tens of millions of these light buckets, and most CCDs now in use don't have this many. The picture you get from them is therefore "grainy," in the same way a fast film is grainy. As technology advances, however, the number of pixels on CCD chips is increasing; some chips now have over a million pixels.

Another problem with CCDs is noise. Fortunately, much of the noise in modern systems can now be removed. What is noise? The easiest way to understand it is through a thought experiment. Assume you take four identical exposures of an object in the sky. During the first exposure you might measure 10,003 photons striking your plate, 9995 during the second, 9997 during the third, and 10,005 during the fourth. The average of them is 10,000 photons. But if you continue to take exposures, only occasionally will you get exactly 10,000 strikes; most of the time it will be slightly different. This randomness is called noise.

Astronomers now realize that there are four sources of noise. First, there's photon noise. To understand it, assume you take a square plate outside when it's raining, and count the raindrops that hit it for 5 minutes. If you compare this number to the number hitting it during the succeeding 5 minutes, assuming the rain stays constant, you will find that the two numbers will be slightly different. You have the same situation with photons hitting a pixel. It can be overcome to some degree by lengthening the exposure.

A second type of noise, called thermal noise, is generated because CCD chips create a small signal even when no light falls on them. It is proportional to temperature, with higher temperatures producing a greater signal. Astronomers cool their detectors to overcome this—sometimes to temperatures as low as –270°C. Taking two exposures also helps. If an exposure is taken of the object and the background, then one of the background itself, and it is subtracted, much of this noise will disappear.

The other two types of noise are called readout noise and quantization noise. Readout noise is caused by the electrons in the system; the amplifier, for example, is not a perfect detector of electrons. The best CCDs are not seriously affected by this. Quantization noise is related to the fact that you are dealing with digital data. Again, it can be minimized, but not eliminated.

The image that the astronomer gets has considerable noise in it, and a lot of work is required to get rid of it. This is usually done later, after the data are collected.

Let's turn now to how astronomers use CCDs and other astronomical equipment.

INSIDE THE UKIRT DOME

The sign taped to the computer terminal read, "Have you hugged your telescope today?" Above it a large window looked out on the huge orange and blue infrared telescope; several instruments were attached to its base. Red numbers flashed on digital displays throughout the control room. Colorful astronomical pictures decorated several of the walls, and on a shelf were the staples for the night: a bag of "Chips Ahoy" cookies, an orange, and two bottles of juice.

An astronomer sat in a large swivel chair studying the screen in front of him. This is where data from the telescope come in. Off to the side was a small TV screen that gave a view of the sky.

It was late afternoon and the night crew would not arrive at the telescope for another couple of hours. Earlier I had talked to

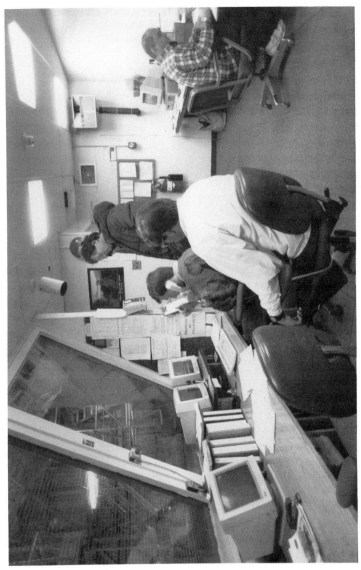

The control room of the United Kingdom infrared telescope.

Colin Aspin.

Colin Aspin, a resident astronomer at the Joint Astronomy Centre in Hilo; he is in this room two or three nights a month.

"I usually go to the summit about 6:00 P.M.," he said. As a resident astronomer his major job is to assist visiting astronomers, but he also has a research program of his own. On most nights there are two or three astronomers in the control room, along with a telescope operator. "The first thing we do when we get to the summit is check over the telescope, and make sure everything is working. Then we go to the instrument on the base of the telescope that we will be using that night and set things up so that light from the telescope enters it." With the UKIRT telescope all instruments are at the Cassegrain focus in the base of the telescope.

Returning to the control room Colin checks the computer

software that he will use that night, making sure it runs without any technical problems. Then he checks the detector. "There are tests you go through to make sure they are behaving—don't have excess noise," he said. Finally when everything is prepared, they wait for darkness, and as soon as the first stars are visible they check the seeing, since the program they have planned for the night may depend on how good it is.

"Once you get started it can be very routine," said Aspin. "You might type in a command to take an exposure, or change a filter, or you might tell the telescope operator to move to a different object."

Aspin's interest in astronomy began in high school. "In the final year we had certain options," he said. "A local amateur astronomer was giving an astronomy course. It was only three or four months long, but I decided to take it. I really enjoyed it—I was hooked after that." He bought a small 3-inch reflector and began observing variable stars in his backyard.

He knew he wanted to become an astronomer, but at Leicester University in England, where he got his bachelor's degree, his adviser told him it was best to combine astronomy with physics, so he took courses in both areas. From Leicester he went to the University of Sussex for a master's degree, then to the University of Glasgow for a Ph.D. His thesis was on the variations of optical polarization in close binary systems. "I didn't do much observing at that time," Aspin said. "It was basically a theoretical thesis." But after he graduated he went to the University of Edinburgh on a research fellowship where he worked on a CCD system. It was tested at several sites, including one in Arizona.

In 1984 he began working for the Royal Observatory of Edinburgh, the institute that directs operations of the Joint Astronomy Centre, and in 1987 he transferred to Hawaii.

I asked him to recall his most memorable night at the Mauna Kea Observatory. He rolled his eyes slightly and leaned back; after a few moments he smiled and said, "It was quite a few years ago. I was involved with building the first infrared camera that was used here. We started on it in 1984 back in Edinburgh. I worked on it right up to 1987. When I came here we took it up to the telescope, pointed it at the Orion nebula, and took the first infrared picture. After working for 5 years on the camera, it was quite a thrill to see that picture."

This was the first time astronomers at UKIRT were able to take infrared "photographs." Before this they had to scan back and forth, creating what is called a mosaic. With an object as large as the Orion nebula, this could take several nights.

IRCAM, as the infrared camera is known, was considered to be an important breakthrough for UKIRT. It was one of the first instruments that allowed astronomers to "see" the sky as it would appear if our eyes were heat sensitive. The infrared detector array consists of 62 by 58 (3596) closely packed infrared sensors which image a small section of the sky. It is built into a cryostat and cooled to between -270 and -200°C.

Aspin is using this camera on two projects at the present time. In the first he is searching for young stars. I asked him about it. He pushed a large colorful print in front of me. "Here's a picture of one of the regions we're interested in," he said. It was an infrared picture of the star cluster GGD 27. "Looking at the infrared is a good way of telling whether some of the objects here are young— born in the gas cloud—or whether they are background objects seen through the cloud." He said he was able to conclude that GGD 27 was a young cluster with at least three very young stars in it. He went on to describe his second project to me, again pushing an infrared picture in my direction. "Until a few years ago this object was thought to be a plain old boring planetary nebula. A colleague of mine, however, found large jets associated with it; material was being shot out of it in two opposite directions. We decided to look at it in the infrared, and found highly collimated infrared jets coming out of the center that didn't fall on the same axis as the optical jet." He paused. "It's an intriguing object."

Much of Aspin's early work was done using IRCAM. IRCAM was, however, replaced in 1994 by an even better infrared camera, IRCAM III, which has an array of 256 by 256 pixels (65,536). The readout noise is much lower, so you can get information faster.

M1−16 − K

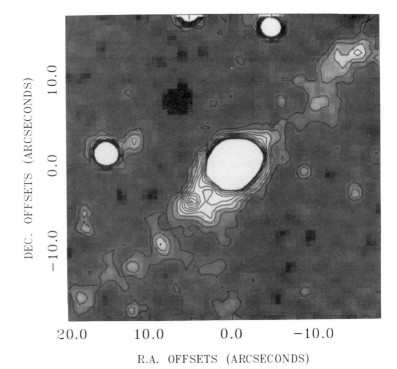

An infrared picture obtained using IRCAM. (Courtesy Colin Aspin)

Aspin estimates that it gives an improvement of 20 over the current system. Arrays of this size are already being used on several of the other telescopes on the mountain.

Although all resident astronomers have research programs of their own, most of the observing done at the observatory is done by visiting astronomers. Astronomers from Canada, France, Great Britain, and other countries fly to Hawaii for a few nights on the telescope. How do they get observing time? It begins with a proposal, which the observer sends to a committee. There are usually requests for about three times as much observing time as there is available on most telescopes, so the committee members rank the proposals and the top one-third get time on the telescope—usually two to four nights. The astronomer then flies to Hawaii and goes to Hale Pohaku.

Kevin Krisciunas of the Joint Astronomy Centre has spent many nights on the summit of Mauna Kea in the UKIRT dome. "With all the checks and calibrations you usually don't get going until about 8:00 P.M.," he said. "There are usually a couple of observers and a telescope operator in the room. One of the two observers runs the instrument, and the other reduces data. The telescope operator points the telescope and advises people on the proper use of the instruments. He, or she, has absolute control of the telescope, so if it starts to fog up and he wants to avoid ice on the telescope, the telescope operator has the power to press the stop button and close the dome in the middle of an observation, no matter how loudly the observer complains." This can be a frustrating experience for the observer. They only have a certain number of nights—usually two to four—to complete their observations, and these days are not always clear. "It's difficult to tell an observer that he has to stop observing," said Krisciunas. "But there are other people waiting for the telescope, and we can't take a chance on damaging it."

One of the most difficult things for astronomers, particularly the first night of their observing run, is staying awake all night. Not only do they have the normal drowsiness associated with changing nights into days, and days into nights, but they are in an atmosphere that contains considerably less oxygen than normal. How do they stay awake? Several things help. First, there are snacks—fruit, crackers, and so on. And there's music. "Music is an important part of observing," said Krisciunas. "In the early part of the evening you can play classical and light music, but as the evening wears on the observers usually switch to something much more lively, such as rock."

By three in the morning many astronomers are usually struggling to stay awake. Four A.M. is even worse. Beyond that, how-ever, they start to wake up. Their state of drowsiness depends, of course, on how well things are going. If emergencies occur, if clouds come in and stop observations, or if the wind gets so high that the operator has to close the observatory, their drowsiness quickly changes to anticipation.

I asked Krisciunas about his favorite night at the observatory. He thought for a moment, then assuming I was asking about important breakthroughs, he said, "Only rarely when you're tak-ing data do you know you're onto something. Occasionally you get some spectra or other data, particularly if it's a new instru-ment, that you get excited about. But just knowing that your instrument is working according to specs, and that the sky is really good is reassuring. The excitement usually comes later when you analyze your data."

DOWN IN SUBMILLIMETER VALLEY

Looking down the slope from the UKIRT you see the dome of the James Clerk Maxwell telescope. The control room here is quite similar to the one at the UKIRT. From it you get a spectacular view of the Maxwell telescope through a large window. The observers sit in swivel chairs in front of this window with TV screens and computers before them.

It was a chilly afternoon in mid-January when I visited the room. Several people were preparing the telescope for the evening run. Somebody sat typing at a keyboard, and off to one side of him was an open bag of cookies. In one corner was a computer with a colorful chessboard displayed on the screen. As I talked to the telescope operator the telephone rang. Nobody moved, and it rang again; finally after several rings somebody grabbed it and began talking in a low voice. Just behind him was a strip chart recorder. I was surprised to see a strip recorder used in this day of digital data and computers.

Later I was down at the midlevel facility talking to Michael Rowan-Robinson of Queen Mary College in England who had an observing run that evening on the Maxwell telescope. Dishes clattered in the kitchen, as a radio blared rock music. A number of other astronomers were seated at other tables around us chatting about their work. How was the seeing? How did you feel last night? Several were getting into detail about their work. To a nonastronomer it would have sounded like another language.

"The Maxwell telescope is unusual in that you don't necessarily have to work at night," said Rowan-Robinson. "If the atmosphere is good you can go up before sunset and start. I usually have dinner about 4:00 P.M., then go up if the weather is good. I start in right away while it's still light."

If the clouds are cirrus they hardly affect JCMT observations, since they are ice crystals. But wet clouds can be a problem: if they pass in front of the telescope while you're taking data, they can be a nuisance.

The setup is roughly the same with the JCMT as with the other telescopes. "The first thing you do when you get to the summit is check over the instruments and get things going," said Rowan-Robinson. "We then find a planet such as Mars or Jupiter. They are important in calibrating our equipment." Once all of the preliminary work is done, observations begin in earnest. "The observations are lengthy scans ... very tedious," said Rowan-Robinson. "When the data comes in we analyze it, trying to determine if we are seeing anything. All the time we're battling noise from the sky and so on. It takes most of the evening to build up a picture of an object."

But just as scanning in the infrared was overcome so too will scanning soon be overcome in the submillimeter region. An instrument called SCUBA is to go into operation at JCMT soon. "When SCUBA becomes operational next year, it will revolutionize submillimeter astronomy," said Ian Robson, director of JCMT. "It will be like going from a single photo cell to a CCD array. For the first time we'll be able to take a picture of a galaxy in the submillimeter region. We'll be able to map a galaxy like M82 in a few minutes, where it now takes the whole night."

As with most telescopes on the mountain, JCMT has a telescope operator; the observer does not touch the telescope itself. At the nearby Caltech submillimeter observatory, however, this is not the case. "We let the astronomers run the telescope and equipment all by themselves," said Anthony Schinckel, the technical manager at CSO. "I or one of my staff go up with new observers. Typically, a new observer will get one night's instruction. We show them right through the system, and tell them about problems that can arise. Then it's up to them." He paused as he shuffled some papers on his desk. "It's really very easy to run—it's 'user-friendly,' as the saying goes. I go up sometimes, though, to help even experienced observers if the project is different, or particularly difficult."

On this night Tom Phillips, the director of CSO, and three Caltech astronomers were at the observatory. Over the next few nights they planned on studying a group of bright colliding galaxies whose central region was undergoing an explosive creation of stars—a "starburst," as they are called. These objects generate a tremendous amount of energy—up to 100 times as much as the Milky Way generates. The group hoped to get a better idea of what was going on during the star creation process.

When they arrived at the dome, the usually clear summit was cloudy and it appeared as if the night's observing might have to be canceled. After sitting and waiting for several hours, however, it cleared up and the telescope was aimed at the starburst galaxy. A number of adjustments were made and data appeared on the screen, indicating they were on the source. Music streamed over the audio system as the night wore on. Further calibrations and adjustments were made, but everything continued to work. By 4:00 A.M. they were beginning to feel exhausted, but it appeared as if the night had been a success. Shortly after daylight they drove down to Hale Pohaku to sleep and start over again the following night.

INSIDE THE CANADA–FRANCE–HAWAII DOME

The control room at the Canada–France–Hawaii dome differs from most of the others only in that the telescope operators are at

Control room of the Canada–France–Hawaii telescope. (Courtesy Canada–France–Hawaii Corporation)

some distance from the observers, and frequently communicate with them via a microphone. Again, astronomers spend the evening sitting before a TV screen and computer monitors. One of these astronomers is Olivier Le Fèvre. As a resident astronomer, Le Fèvre's major responsibility is helping visiting astronomers. He shows them how to use the instruments, how to take data, and he helps set up the instruments for them, but he also manages to squeeze in considerable time for his own research. His research interests are cosmology and extragalactic astronomy, but he has done a considerable amount of work recently on gravitational lenses. Gravitational lensing occurs when the light from a distant object such as a quasar or galaxy is bent or broken up by an object along the line of sight.

"Most of the early work for a run is done during the day," said Le Fèvre. "You have to set up all the instruments you will need for the observing run. So it sometimes takes a day or two of hard work setting up the spectrograph, or whatever you're going to use. Once that's done you can relax." As we saw earlier, the Canada–France–Hawaii telescope can be used in three different configurations: prime focus, Cassegrain focus for visible or infrared, and Coude focus. In each case, physical changes have to be made to the telescope. If the telescope has to be changed it can take several hours.

Born and raised in Paris, Le Fèvre got a bachelor's degree from the University of Paris and a Ph.D. at the University of Toulouse, where he worked on clusters of galaxies.

"I knew from early on I wanted to be an astronomer," he said. "I began as an amateur astronomer. Several of us dug an old refractor out of the attic when we were in high school. We spent a lot of time looking through it at night. Later I began building telescopes—grinding mirrors. I built several 8-inch reflectors."

He began by building a telescope for himself, then found that several of his friends wanted one. "Grinding mirrors was a good source of income in college," he said.

His parents, both of whom were in graphic arts, were surprised when he told them he wanted to become an astronomer.

"At first they looked at me with a 'weird eye,'" he said. "They knew little about science, and even less about astronomy, but when they saw how excited I was about it they began to encourage me. So it worked out well in the end."

Le Fèvre's description of a night at the observatory differed little from what I had heard previously. "You just sit in front of a computer screen and take data and analyze the results. The picture you get from the telescope is in raw form—it's a rough image. You have to process it with the computer to extract information. There's quite a bit of noise you have to remove." He leaned back and smiled. "You spend the night at the observatory taking data, and you have enough for six months' work. You don't even find out for quite a while if you've been successful."

Another resident astronomer at the CFH facility is François Rigaut. He is also from France, born and raised near Paris. He started off in physics, with little interest in astronomy, but several of the physics courses he took brought him in contact with astronomy. It intrigued him so much that he decided to major in astronomy. His first experience was with a 4-inch reflector.

Rigaut received a bachelor's degree in physics and math from the University of Lyons in France. He got a Ph.D. from the University of Paris where he did considerable observing at the Paris observatory. In addition to helping visiting astronomers at CFHT, Rigaut is working on a new adaptive optics system for the telescope. His Ph.D. thesis project was connected with an adaptive optics system called COME-ON. Adaptive optics, as we saw earlier, is a technique for improving the resolution of the telescope. The astronomer analyzes the wave front going into the telescope and corrects any distortions in it.

The first adaptive optics system was built in the United States for military applications. The group Rigaut was working with wanted to adapt it to astronomy. They started on a prototype in 1988, completing it in 1989, and after testing it in France, and later in Chile, they were pleased with its success.

Rigaut is using the information and experience he gained on this project to help get a new adaptive optics system for the CFHT.

François Rigaut.

"I haven't been up to the summit very much," he said. "I've only been here for four months and I've been busy with the adaptive optics system. It will probably be a couple of years before it is completed, then I'll be up there a lot more."

The University of Hawaii is also working on an adaptive optics system that they will use on the CFHT and the 88-inch telescope.

REMOTE OBSERVING

Remote observing—that is, sitting at a console miles, or perhaps thousands of miles, from the telescope and collecting data—is becoming increasingly common on the mountain. When the Keck telescope is finished, most of the observing with it will be remote observing.

Most of the remote observing on the mountain so far has been done with the UKIRT, although some has been done using the JCMT and also the CFHT. There are several remote sites in the United Kingdom that can be linked to the UKIRT. One of the major ones is at the Royal Observatory in Edinburgh.

Remote observing can be passive or active. In passive observing the observer watches the progress of the data coming in but does not control any of the observations, and usually has little input. In active observing the remote observer controls some of the observations, makes decisions, and is directly involved with the overall process. Most remote observing has been active observing.

Caroline Crawford of the Institute of Astronomy at Cambridge, England, was a UKIRT remote observer at the Royal Observatory in Edinburgh (ROE). She wrote about the experience. "I was given a quiet room at ROE with a couple of workstations. One workstation was used to display the continually updated telescope status, instrument status, and control screens—basically the information available to the observer at the telescope. The second workstation enabled fast reduction and display of our data using any of the standard software packages available. There was another terminal for e-mail to the observer in Hawaii, logging on at UKIRT, copying data files, and there were two outside phone lines. Data were accessible as soon as they were written to the disk at UKIRT. I could pull fresh IRCAM mosaics to the ROE [screen] in a few minutes."

Crawford said she encountered a few minor hitches during data reduction and analysis, but she admitted they might have been a result of her unfamiliarity with the system. "The problems

I encountered were comparatively trivial, and I was pleasantly surprised with the system," she said.

The most positive aspect of the experience, she said, was that you definitely feel involved in the observing. "I was able to give direct feedback about signal-to-noise ratios, possible saturation of images, estimates of further time needed. . . . Each night I was able to spot something that would otherwise have cost us some time. Most importantly, I felt like an equal partner in the observations taken, and feel that their quality has been substantially improved by my involvement."

She admitted that she started with a skeptical view but was delighted with the results.

CHAPTER 8

Running the Observatory: The Directors

When Gerard Kuiper flew over Mauna Kea 30 years ago, looking down on the bald, desolate summit as a possible observatory site, he could hardly have visualized what we see there today. Not only has the observatory grown tremendously in the last few years, but with current projects, it will continue to grow for several more years. Strangely, though, it is different from most other observatory complexes. Kitt Peak National Observatory in Arizona, and the European Southern Observatory (ESO) in Chile both have many telescopes, but each is under a single administration. The observatories on Mauna Kea are not; they are separate research centers, and are administered separately. "It could be described as a research park," said Bob McLaren, associate director for Mauna Kea of the Institute for Astronomy in Honolulu.

The offices of the observatories are also not only separate but widely separated geographically. The offices of the Institute for Astronomy, along with other major facilities, are in the Honolulu suburb of Manoa, at the north end of the University of Hawaii campus. The Joint Astronomy Centre, which oversees UKIRT and JCMT, has offices in Hilo, and Keck Observatory and the Canada–France–Hawaii Telescope have offices in the northern part of the Big Island in Waimea. Astronomers from the various observatories stay together at the midlevel facility at Hale Pohaku, and frequently share instruments and ideas, but they do not work under the same management. The major thing uniting them is the University of Hawaii, which, as leaseholder and manager of the observatory site, is part of them all.

THE INSTITUTE FOR ASTRONOMY

It was pouring rain as I stepped off the plane in Honolulu on my way to the Institute for Astronomy (IFA). By the time I had struggled through the heavy early morning traffic up to Manoa, however, the rain had stopped and the sun was shining.

The Institute for Astronomy, which is part of the University of Hawaii, is housed in a large, two-story concrete building, with lush green hills surrounding it. The associate director for Mauna Kea, Bob McLaren, stood and shook my hand as I was ushered into his office. His office was large, with windows along one side, and photographs of the observatories and Hawaiian scenery decorating several of the walls.

The institute now has more than 200 full-time employees and an annual budget of over $15 million. "Mauna Kea is a young observatory, as observatories go," said McLaren. "We've only been on the mountain for 30 years." He went on to tell me about the University of Hawaii's role in overseeing the observatories. "Most of the land above 12,000 feet on Mauna Kea has been designated a science reserve. It is leased to the University of Hawaii on a 65-year lease. The University is therefore the leaseholder, and it, in turn, gives the various observatories subleases in return for observing time on their telescopes."

The University of Hawaii, in turn, McLaren explained, has several responsibilities. First, it is the go-between with the state, which actually owns the land. Second, it is the guardian of site quality, and it is responsible for making sure that good seeing is maintained, with no interference from excess lighting or stray radio frequency radiation. Third, it looks after the road to the summit and maintains the midlevel facility at Hale Pohaku. As we saw earlier, the upper four miles of the road have been paved to

The Institute for Astronomy at Manoa.

keep the dust down. I asked McLaren if there were plans to pave the lower four miles. "It will eventually get done," he said, "but it may not be before the year 2000."

He talked at length about the accomplishments of various people at the Institute for Astronomy. Then, pausing, he grinned. "We enjoy a big advantage over most astronomy institutions," he said. "We have access to all the telescopes on a mountain with the best seeing in the world."

McLaren came to the institute in 1990. Canadian by birth, he took all of his degrees, including his Ph.D., at the University of Toronto, and like many astronomers he started out in physics. His Ph.D. thesis, which he completed in 1973, was an attempt to learn about molecular forces in rare gases by measuring the elastic constant of solid neon using lasers.

I asked him how he got into astronomy. "When I got my Ph.D.," he said, "there were only two ways for me to go in the laser field. I could go into an applications area such as electro-optics, and I wasn't interested in that. Or I could go further into molecular physics—measuring the properties of condensed matter and interacting molecules. And I wasn't really interested in that. So I decided to do something different." He leaned back in his chair, then with a chuckle added, "I had also been in a basement lab for five years and had had enough of that."

He applied for a NATO postdoctoral and used it to go to the University of California at Berkeley to work with Nobel laureate Charles Townes. In California he worked on an infrared airborne project using a Lear jet, and also on spectroscopy applied to astronomy. This was his first contact with astronomy and he decided to stay with it.

In 1975 a position in astronomy at the University of Toronto opened up, and McLaren applied for it. He was selected, and for the next seven years he taught and did research in Toronto. In 1982 he took a sabbatical to the Canada–France–Hawaii Telescope, and while there, a resident astronomer position became vacant. The director asked him if he was interested in it. It didn't take him long to decide; the long, cold, winters of Toronto had taken their toll,

Bob McLaren.

and Hawaii seemed like a welcome change. He was soon associate director, then director, and in 1990 he moved to the Institute for Astronomy as associate director for Mauna Kea.

In the office next to McLaren's is Don Hall, the director of the institute. He came to the institute in 1984. Born in Sydney, Australia, Hall received his B.Sc. in physics from the University of

Sydney in 1966. He then went to Harvard University, receiving a Ph.D. in 1970.

Upon graduation Hall went to Kitt Peak National Observatory in Arizona where he worked on instrumentation for both the solar telescope and the large 4-meter telescope. His major interest until then had been the sun, but he now shifted to star formations, variable stars, and the core of our galaxy and others. In 1982 he moved to Baltimore to become deputy director of NASA's Space Telescope Science Institute. While there he helped establish the Space Telescope Institute into a first-rate science institution. He was also responsible for setting up much of the machinery needed for conducting the science program once the Hubble telescope was launched.

In 1984 Hall came to the Institute for Astronomy as director. In the nine years he has been there the institute has grown from 130 to over 200 full-time employees.

THE JOINT ASTRONOMY CENTRE

Above the university, overlooking the city of Hilo, is the Joint Astronomy Centre. It is a large, two-story red building and employs about 60 people. The day-to-day operation of the United Kingdom infrared and JCM telescopes is directed from here. Resident astronomers and support people work here when not on the mountain. The Joint Astronomy Centre is administered from the Royal Observatory in Edinburgh, and from offices in Canada and the Netherlands.

The director of the Centre when I was there was Malcolm Smith. Ebullient, slightly built with curly gray hair, Smith is the author of over 100 scientific papers. He says he was determined from a very young age to become an astronomer. Born in Devonshire, England, his interest in astronomy began when he was about 8. The exploits of Dan Dare, a fictional character in one of the local papers, captivated him. Dare visited many strange places—Mars, Venus, Jupiter. Smith had never heard of them so he pulled

The Joint Astronomy Centre in Hilo.

Malcolm Smith.

an old family encyclopedia from the shelf, dusted it off, and looked them up. Soon he was reading everything he could about astronomy, and by the time he was 10 he had decided to become an astronomer.

When he told people about his ambition, however, many of them thought he was a little crazy. Even in high school when he told his career advising counselor that he wanted to become an astronomer, he was discouraged. "Young man," the counselor said, "be sensible, you are good at math. You could get an excellent position in the bank."

Smith, however, was not deterred. Still, he was not entirely sure what astronomers did; he wanted to meet an astronomer and talk to him, but in the town where he lived there were no astronomers, not even amateur astronomers. His physics teacher was the closest thing there was to an astronomer, and was one of the few people who encouraged him. He even gave Smith private lessons in physics. "He taught me optics that was very useful to me later on," said Smith. Physics was, of course, important as a foundation for astronomy and Smith was advised to start out with it, so when he went to King's College, London University in 1961, he majored in physics.

One of the senior members of the x-ray crystallography group at King's College was Nobel laureate Maurice Wilkins. According to Smith, his association with this group allowed him to get a summer research fellowship at the Department of Biophysics at Roswell Park Memorial Institute in Buffalo, New York. It was his first visit to the United States, and although he was working in x-ray crystallography, his ambition was still to become an astronomer. So while he was in Buffalo he made a trip to Chicago to take in a planetarium show. After the show he walked down the long corridor of the planetarium offices and knocked on one of the doors. As it turned out, it was the director's office. The director invited him in, told him to sit down, and they had a long talk. Smith told the director of his interest in astronomy and his ambition to meet an astronomer; he also mentioned that on his way back to England he would be stopping in California. The director smiled and said, "If you're going to California and want to visit the large observatories, and meet some astronomers, I'll write them telling them you are coming. All you'll have to do is give them my card when you get there." He gave Smith one of his cards.

When Smith got to California late that summer he went to Santa Barbara Street in Pasadena to the headquarters of the Palomar Observatories and presented the card. "I was introduced to this gray-haired man named Ira Bowen, who I had never heard of," said Smith; with a chuckle he added, "That helped a lot at the time, because I later found out he was the director." Smith talked with Bowen for almost an hour. Bowen showed him some of the plates from the telescopes, and Smith in turn told him about his ambition to become an astronomer.

Smith finished his bachelor's degree in physics in 1964, then went to University of Manchester in England for a Ph.D. in astronomy. His thesis was on observational instrumentation. "I did a lot of instrumental work on optical interferometers of various kinds," said Smith. "My Ph.D. project was to build an interferometer. One of the faculty members, Franz Kahn, was particularly helpful; he was good with students, and had connections. He said to me one day, 'Would you like to work at Palomar, Lick, or Kitt Peak?' I was amazed; it wasn't easy to get appointments at any of those places."

Smith went to Kitt Peak when he graduated in 1967. Just as he had built an interferometer for his Ph.D., he was to design and build one for use on the telescopes at Kitt Peak. He had a one-year appointment, and spent the first 8 months of it designing the instrument. The year was almost up when he started constructing it. One day he noticed a note on one of his supervisor's desk saying, "This guy Malcolm Smith has been here for almost a year. What has he done? Has he finished his project?"

Smith worried about his predicament, but wasn't sure what to do. Earlier, however, he had met Bart Bok of the University of Arizona, and Steward Observatory. Bok had seemed friendly so Smith went over to Steward and told him about his problem. "Let me come over and see what you have done," Bok said. Bok looked over the design and gave him some advice. From then on he came to Kitt Peak about every two weeks and looked over what Smith had done. "He gave me a lot of encouragement, when other people were severely criticizing me," said Smith. "It was comforting to have his support."

When Smith completed the interferometer he took it down to Cerro Tololo in Chile where he had been offered a position as assistant astronomer in 1969. "It was a great place," he said. "I did much of my most exciting research there . . . mainly because I didn't have many administrative duties." He stayed in Chile until 1976, when he went to the Anglo–Australian Observatory in Australia.

In 1979 he returned to Edinburgh, Scotland, as head of technology, where he remained until 1985 when he came to Hawaii. During his career he has worked in x-ray, optical, infrared, and submillimeter astronomy. Although he doesn't get nearly as much time for research as he once did, he's still involved in several projects. Gaseous regions around quasars are one of his main interests. "We've been investigating these regions using infrared," said Smith. "It's only recently that anyone has had instruments with enough sensitivity to study them in the infrared." His interest in these gaseous regions is whether they are falling into the quasar, or being ejected from them. "It may seem straightforward, but it's actually a tricky measurement," he said.

In the office next to Smith's at the Joint Astronomy Centre is the director of JCMT, Ian Robson. He has been at the center since 1992. Born in northern England, Robson's interest in astronomy, like Smith's, began when he was young. He would lie in bed at night looking out through the window in his room at the stars, wondering why they were arranged in such strange patterns. One day he saw an astronomy program on television narrated by Patrick Moore. Robson decided to write Moore, asking him about the patterns he was seeing; he described one that particularly intrigued him. Moore wrote back telling him it was the constellation Orion. He also recommended some books where Robson could learn more about astronomy. Robson was soon fascinated with astronomy, and by the time he was 11 he had a refractor and was an avid amateur astronomer. From his backyard he studied the moon, planets, and stars. He continued to read every book he could find on astronomy. "It was fun learning so much about what was in the sky," he said.

Strangely, though, despite his intense interest in astronomy, he was not planning on becoming an astronomer. For several years his ambition was to fly planes—to be a fighter pilot. But when he was about 15, his eyesight began to deteriorate, and he knew he wouldn't be able to get in the air force, so he decided to become an astronomer.

In 1965 Robson entered Queen Mary College in London, where he majored in physics. He obtained a bachelor's degree in 1969, and stayed on for a Ph.D. in astronomy. A few years earlier the cosmic background radiation had been discovered by Arno Penzias and Robert Wilson of Bell Labs. This is radiation that was left over from the big bang explosion that created the universe. When created, it had a temperature of approximately 3000 K, but it had now cooled to 3 K. In the late 1960s there was still some uncertainty as to whether this radiation was really from the big bang explosion. If it was, its temperature curve had to have a characteristic shape. If you plotted the intensity of the radiation versus frequency, the curve had to increase, then drop off beyond a certain frequency. In other words it had to turn over. But all of the points that had been measured were on one side of this turn-over. It was crucial, therefore, to obtain points beyond it. There was, however, a problem: our atmosphere absorbed the radiation in the region of the turnover, and you had to get above it if you were to obtain these points. Rockets or balloons were therefore needed.

The group that Robson worked with was trying to verify the turnover using a balloon. The experiment was to form the basis of Robson's Ph.D. thesis. The balloon was flown in 1972, and at 6000 feet it burst. "It was traumatic . . . the ruptured balloon and pay-load crashed to the ground. It delayed my thesis, and meant that I couldn't graduate for another year," said Robson. Despite the setback, the group started again, and seven months later they were ready for a second launch. "That second launch was one of the most exciting, and most nerve-racking, moments of my career. I was relieved when everything went well, and data started to come in. We proved that the curve did, indeed, turn over." Robson received his Ph.D. in 1973.

Robson stayed on at Queen Mary and did a postdoctoral after receiving his Ph.D., then was on the faculty as a lecturer for a couple of years. In 1978 he went to Lancashire Polytechnic (now

University of Central Lancashire) in Preston where he started off as a senior lecturer, worked his way up to professor, then finally to department head in 1986.

Robson's association with Mauna Kea began in the early 1970s when he began observing on the 88-inch University of Hawaii telescope. Later he was on the team that designed the United Kingdom infrared telescope. It was during one of his early trips to the mountain, in about 1975, that he found out how dangerous it can be. A blizzard had hit that day, and there were several feet of snow at the summit. Despite this, several French astronomers wanted to go to the Canada–France–Hawaii dome. Robson wanted to get some equipment out of the 88-inch telescope dome, so he decided to go along with them.

A snow caterpillar took them as far as it could go. They had heavy parkas, hats, and gloves on, so when it finally bogged down they decided to hike the rest of the way to the top on foot. Robson, a colleague, the driver of the cat, and four French astronomers started up the remaining two switchbacks to the top. The snow was 3 to 5 feet deep, and it was late in the afternoon.

Within a short distance two of the French astronomers started to get sick; they hadn't acclimatized properly at Hale Pohaku. The cat operator offered to take them back to the cat. The others continued on for some distance until finally the remaining two French astronomers felt too weak to continue and Robson's colleague escorted them back. Robson thought about going back, but was sure they were extremely close to the top so he decided to continue on. It was getting dark now, but he had a flashlight.

With each step it got darker, until finally it was so black he couldn't see a thing without the flashlight. To make things worse the snowdrifts were even higher here, and he knew there were steep drop-offs in the vicinity.

"You couldn't walk," said Robson. "You literally had to fight your way through the huge snowdrifts." Soon his flashlight began to dim, then it died completely. It was so dark he now had no idea which direction to go, but he knew if he kept going uphill he would eventually get to the 88-inch telescope dome. So in the darkness he zigzagged back and forth, making sure he was always going uphill. Finally, he crawled to the back of the 88-inch telescope dome. "I didn't have the strength to make it to the front door," he said. The wind was blowing hard, and it was still snowing, but he knew there was a back door, so he felt around and finally managed to find it, and get in.

"I headed for the shower the moment I got inside," he said. "I stood there for a quarter of an hour trying to get heated up. Then I phoned down to Hale Pohaku to tell them I was all right." They told him that one of his colleagues had gotten worried and had come after him. A little while later he arrived, in about the same shape Robson had arrived in.

"That's something I never want to go through again," said Robson.

In November, 1992 Robson came to Hawaii as director of JCMT. He has two research interests: the study of star formation regions using the infrared and submillimeter, and quasars. His major interest at the present time is blazars—active galaxies that are closely related to quasars. The material around a quasar nucleus forms a spiraling disk called an accretion disk; the material flows inward via this disk to the black hole at the center. It is then ejected at very high speeds—0.9999c—in two beams perpendicular to the disk. If the beam is pointed toward you, you have an optically variable quasar, or B L Lac object (also called a "blazar"). If the beam is in the plane of the sky you would call it a double-lobed radio source.

Robson has been using JCMT and other telescopes to explore these objects, trying to determine the mechanism of the jet and the velocity of the ejected material. One of his favorite objects is the quasar 3C-273—the second quasar found. "We found a flux that varied by a factor of 2 in 3C-273 from one night to the next a few years ago," he said. "I drove down the mountain immediately . . . all excited . . . and sent an IAU telegram announcing the discovery." He sat for a few moments, then smiling, said, "As I grow older I get less excited about such things . . . individual results like that. But I am getting excited about SCUBA." SCUBA, as we noted earlier, will allow astronomers to take pictures in the submillimeter region.

THE CANADA–FRANCE–HAWAII CORPORATION

The headquarters of the Canada–France–Hawaii telescope is located in Waimea, in the northern part of the island of Hawaii. It has a permanent staff of approximately 50, and is administered by the National Research Council of Canada, the French National Center for Science Research, and the University of Hawaii.

A large blue and gray building set back from the street, it was completed in 1982, with an addition being made in 1986. Inside the large front doors is a lounge, or waiting area, decorated with astronomical photographs. A model of the CFH telescope sits on display. Several historical photographs hang on another wall, showing the telescope while it was being built. The interior of the building is taken up by a large open courtyard that contains several small trees and bushes.

The directorship of the CFH Corporation is rotated every three years. While I was there the director was Guy Monnet. I talked to him about the observatory as hammers and saws sounded in the distance. A large business building was going up next door.

I asked him about some of the discoveries made with the CFH telescope in the last few years. "I like to divide discoveries into three classes," he replied. "Discoveries of the first kind are those that would add a new chapter to a book on astronomy. Discoveries of the second kind would add a new paragraph, and discoveries of the third kind, a sentence." He looked out the window for a moment, then turned back. "I feel there has only been one discovery of the first kind at CFH in the last ten years, and that was the discovery of gravitational arcs. The first one was discovered here."

As we have seen, gravitational lensing is caused when the gravitational field of one object, usually a galaxy, breaks up the

Canada–France–Hawaii headquarters in Waimea.

Guy Monnet.

light from an object behind it. You frequently get two or more images of the object. If the alignment is almost exact, however, you get arcs. In theory, if there were exact alignment, you would get an Einstein ring—an enlarged view of the distant object with a dark circular region in the center. Several gravitational arcs have been discovered at the CFH Observatory.

Monnet went on to say he believed that several discoveries of the second and third kinds had also been made. He felt that the work that had been done on massive black holes fell in these categories.

Born in Lyons, France, Monnet went to a Polytechnic in Lyons for his bachelor's degree, then to the University of Marseilles for his Ph.D., where he worked on an instrument for measuring the velocity of gas in galaxies. He stayed on at Marseilles for 14 years, first at the Space Laboratory, then later he transferred to the Marseilles Observatory, where he eventually became director. He then went back to the University of Lyons, and while there his interest changed from the velocity of gas in galaxies to the velocity of stars in galaxies. He left there in 1987 to come to Hawaii.

His current research interest is the dynamics of stars in the center of galaxies. He's particularly interested in the possibility that some galaxies might have massive black holes at their center. "We've examined the center 1 to 2 arc seconds of several galaxies now, including M32 [the small elliptical companion of M31 (the Andromeda galaxy)] and the Sombrero galaxy, hoping to find a massive black hole," he said. He does this by determining the speeds of stars in this region. By determining the speeds of stars close to the center you can determine the mass inside them. If this region is sufficiently small, and the mass sufficiently large, it may be a black hole.

Assisting Monnet in administering the observatory is the associate director, John Glaspey. Soft-spoken, slender, with gray hair and mustache, Glaspey had been at the observatory 4 1/2 years when I talked to him. His office had dark walls, with the traditional pictures of observatories and astronomical objects on them.

I began by asking him what he felt the major role of the CFH Observatory was. "We're a service observatory," he replied. "That means that we have several astronomers on staff here, but their primary role is to support the operation of the telescope, and their own research is secondary. So the whole facility is geared to providing top-quality instrumentation for the Canadian and French astronomical community and astronomers from the University of Hawaii."

Glaspey went on to talk about the uniqueness of the telescope and the seeing on Mauna Kea. "The location of the CFH dome on the mountain is one of the best there is," he said. "We get image quality that other observatories can't approach. The Einstein

John Glaspey.

Cross, for example, has four images. Their total separation is 1.5 arc seconds, and at Kitt Peak the median seeing is 1.3 arc seconds. So there aren't many nights they could actually resolve something like the Einstein Cross. Whereas here, we can easily resolve it any night. We frequently use it to judge the seeing."

Glaspey's interest in astronomy began during a visit to his aunt and uncle on Long Island. They had a neighbor, an avid

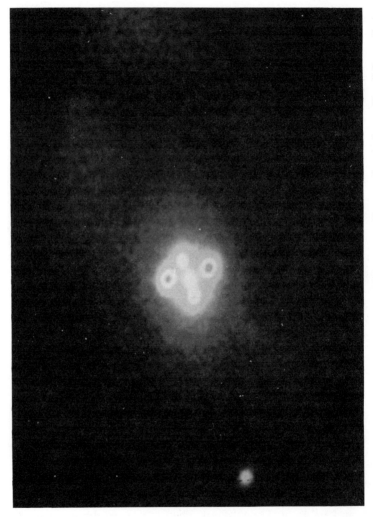

The Einstein Cross as photographed through the CFH telescope. (Courtesy Canada–France–Hawaii Corporation)

amateur astronomer, who used to set up a large refractor on his lawn in the evenings and let the neighborhood kids look at the planets and stars. Glaspey was intrigued with what he saw, and he never forgot it. Through junior and senior high school he did a lot of reading about astronomy—much of it, he admitted, was beyond him. Nevertheless, he was inspired, and soon decided to become an astronomer. "I chose my undergraduate program to prepare me for astronomy," he said. "I had this tunnel vision and never thought of deviating from it."

Glaspey chose Case Institute of Technology for his undergraduate education, a decision he never regretted. "There was this beautiful old building with ivy-covered walls," he said. "The dome of the 36-inch Cassegrain telescope was in the center. There was a lot of fine old equipment there. The observatory had been set up just after World War I, in the early 1920s. So when I was there I got to work with some fascinating equipment—ancient by today's standards—but fascinating nevertheless."

He remembered a two-semester course on practical astronomy in which he used some novel equipment. "It was fun and a great experience for me, one that I appreciate now. Of course then it was a lot of work."

He chuckled when I asked him about his current research. "It's getting tougher every day to do research with all the administrative duties I have." He thought for a few moments. "My most recent project, which has been going on for several years, is a search for lithium in stars. We've been looking lately at blue stragglers [stars that are bluer and brighter than they should be for their age]." Lithium is important because it was produced in small quantities in the big bang explosion. Glaspey and his colleague C. J. Prichet of the University of Victoria found that there was no lithium in the blue stragglers of the galactic cluster M67. This was a surprising result, and one that stumped them at first. They now believe, however, that they have an explanation. In a recently published paper they stated that they are sure large currents are driving the lithium from the surface of the stars into the hot interior where it is destroyed.

KECK

The headquarters of Keck Observatory is in Waimea, just down the street from CFH headquarters. A large white structure with a distinct Western design, it fits in well with the town—the home of the famous Parker Ranch. As you approach the building the first thing you notice is the skylight over the entryway. Inside is a model of the Keck telescope in a glass case. Beyond it is a beautiful courtyard containing trees, lawn, and walkways, in the center of which is a section of lawn that is an exact replica of the Keck mirror.

Well-lit, spacious offices completely surround the courtyard. The area behind the courtyard contains the control room for the telescope. Most observers will take data here rather than at the control room at the telescope itself.

Behind the main building are the sleeping quarters for the visiting astronomers. They will be headquartered here rather than at Hale Pohaku since they will be using the telescope from here. The groundbreaking ceremony for this building took place while I was there, along with the traditional Hawaiian ground blessing.

Unlike the other observatories on the island, Keck does not have a director. The main reason, no doubt, is that it is not finished. Gerald Smith, as project manager, is the senior person in charge. With the completion of Keck I, however, he is now devoting most of his time to Keck II.

"Keck II will be much easier," Smith said. "We kept the Keck I facilities on line so we're able to go immediately to Keck II. We'll have Keck II completed in 3 years—that will be 2 years shorter than the time it took for Keck I."

Raised in California, Smith attended high school there, then spent four years in the air force as a radar maintenance technician during the Korean war. After the war he went to USC where he majored in electrical engineering. He continued on and obtained a master's degree at USC in 1960.

While still in college he began working in the aerospace industry, mainly in electronics and electro-optics. So it was perhaps

Keck headquarters in Waimea.

Inside Keck headquarters showing model of the Keck telescope. Courtyard is seen in the background.

natural that he went to the Jet Propulsion Laboratory (JPL) in Pasadena after he graduated. At JPL he worked on instruments for the NASA satellites that were being launched into space. For the next several years he was building detectors, and managing instrument construction for the planetary spacecrafts Mariner, Ranger, and Surveyor.

"I had a lot of experience with planetary astronomy from spacecrafts before I got mixed up with ground-based astronomy," said Smith. His first experience with ground-based astronomy

Gerald Smith on the right, Ron Laub on the left.

came in 1975. He was working for JPL when he was asked by NASA to oversee the building of the NASA infrared telescope in Hawaii.

When the NASA infrared telescope was completed in 1979, he went back to JPL where he became manager (in 1981) for the Infrared satellite project IRAS. "It looked like it would be a lot of fun," he said. "There would be lots of travel to England, the Netherlands, and so on." But problems surfaced almost immediately. There were problems with the cryostat, squabbling between the various groups, and delays. "We spent a difficult three years working out all the problems. We eventually got the telescope performance up to where we thought it would be satisfactory, but it was a traumatic three years. It was one of those projects where everything was very painful right up to the end. Then we launched it, and it worked so well, no one remembered the problems. They just thought it was a tremendously successful project."

With the completion of IRAS, Smith came to the University of California-Caltech 10-meter telescope project. He had worked on it part-time earlier; he now came on full-time. The project was funded by Keck shortly thereafter. I asked him to compare the Keck and IRAS projects. "There was very little dissension at Keck," he said. "It was a challenging and cooperative effort, and it has been a rewarding one."

As Smith has turned his attention to Keck II, the responsibility for Keck I has been given to Peter Gillingham. He had been there only a few months when I talked to him. Tall, with a broad smiling face, Gillingham talked excitedly about his new position. He is operations director of Keck I, with the responsibility of overseeing its completion, commissioning the instruments that will be used with it, and developing plans for operating the telescope.

Born in Brisbane, Australia, Gillingham went to the University of Queensland in Brisbane where he got a degree in mechanical engineering. After graduation he worked on defense research for a few years, then got a job at the Greenwich Observatory in England. That was his first taste of astronomy. He worked on instrument development, integrating electronic image detectors and spectrographs. He also participated in the completion and commission of the Isaac Newton telescope in 1967. Its 2.5-meter mirror was the largest in Britain at the time.

At the time, Britain was also a partner in the 4-meter Anglo-Australian telescope being built in Australia. Gillingham got a job

Peter Gillingham.

with it, and went back to Australia. And for 12 years he was officer in charge of the telescope.

In 1991 he came to Hawaii to look at the UKIRT and JCMT. During this time he paid a brief visit to the Keck telescope, never dreaming he would be working on it one day. "I thought it was an exciting project," he said. "I was very impressed. There were about 14 out of the 36 mirrors in place." When he got back to Australia he saw an advertisement for a job at Keck and applied for it, and got it.

Gillingham said he spends most of his time at the headquarters in Waimea, getting up to the observatory about one or two days a week. Much of his time is taken up with meetings and scheduling.

He is sure the Keck telescope will perform up to expectations. "When the seeing is good we expect Keck to produce images you could not get anywhere. It should win hands down over any other telescope."

And it does appear to be living up to expectations. In late March 1993 astronomers at Keck announced they had photographed the most distant galaxy ever seen, a galaxy 12 billion light-years away. It took a half hour to get the infrared image.

Telescopes are only successful, however, if they are used successfully. Let's turn now to the research, and the discoveries that have been made with the telescopes on Mauna Kea.

CHAPTER 9

Monster at the Core

Research at the Mauna Kea observatories is diverse. In one of the domes you might find a group of astronomers searching for a comet, in another a group working on galaxies, and in a third astronomers stretching the limit of their telescope, exploring objects near the edge of the universe. At one time much of the research was directed at planets, moons, and stars. In recent years, however, a considerable amount of attention has been paid to galaxies and other extragalactic objects. As more and more has been learned about galaxies, in fact, it has become increasingly clear that the machinery that runs them is a mystery of the first degree. Deep in the heart of a galaxy is a mechanism that produces prodigious amounts of energy. In some galaxies, referred to as active galaxies, energy is literally pouring out of the core.

What produces this energy? Many astronomers believe that a bizarre, massive monster referred to as a black hole may be responsible. Black holes have such a strong gravitational field that nothing, including light, can escape from them. John Michell of Yorkshire, England, postulated their existence back in 1784, but few astronomers heard of them until 1939 when Robert Oppenheimer and his student Hartland Snyder of the University of California at Berkeley applied Einstein's theory of relativity to the collapse of a star. They knew that when the nuclear furnace of a star went out it would be overwhelmed by gravity, but what would eventually happen to the star? Einstein's theory showed that if it was sufficiently massive it would collapse and could end as a hollow black sphere a few miles across. All of the mass of the star would be at the center of this sphere—this black hole—but if you passed through its surface in search of it you could never get out. Once inside there was no escape.

At the time of Oppenheimer's discovery they were not called black holes. This name wasn't applied until the late 1940s. Nevertheless, they were intriguing. And yet Oppenheimer published only one paper on them; World War II broke out and he was soon overseeing the building of the atomic bomb. After the war he never returned to the project, and for many years there was little interest in black holes. The reason, no doubt, was that there was no proof that they existed; they existed only on paper. No one had ever seen anything resembling a black hole in the sky. They were only a few miles across, and astronomers wouldn't have been able to see them directly anyway, but there was also no indirect evidence for them.

The war put a halt to astronomical research in 1941, and many astronomers were diverted to wartime projects. Few had heard of black holes, and the few who had never gave them a second thought. When the war was over they went back to their observatories and initiated new research projects. Many of them now had a strong background in radar and electronics, so radio astronomy soon began to flourish, particularly in England and Australia.

Radio astronomers began cataloguing the objects in the sky that gave off radio waves. By the early 1960s they had noticed that a number of these radio sources were different from the others. Most were extended sources such as galaxies, and could easily be seen with optical telescopes, but some of them were sharp, point-like sources, and they could not be seen. Astronomers began applying ingenious tricks to narrow in on them, and in time they managed to isolate a couple. To their surprise they found that they looked like stars. And when their light was passed through a spectroscope so that the various frequencies could be analyzed they discovered that they were exceedingly distant objects. They were out beyond the farthest galaxies; furthermore, they were incredibly energetic, pouring out more energy than an entire galaxy. It seemed impossible. How could a stellarlike object give off more energy than a hundred billion stars (the number in a galaxy). Nothing like this had ever been seen before.

The objects were soon called quasars, and for several years they were a major astronomical mystery. Theoreticians struggled to explain their energy output. Nuclear energy was out of the question; it was far too inefficient. But there were black holes; under the proper conditions they could produce tremendous amounts of energy. Theoreticians soon zeroed in on them.

The black holes generating this energy, however, would have to be much more massive than stellar collapse black holes—billions of times more massive. Astronomers therefore began referring to them as supermassive or just massive black holes.

How do black holes generate energy? A black hole all by itself would not generate energy; fuel is needed, and this fuel comes from nearby stars and gas. If a star, for example, passed sufficiently close to a massive black hole, it would be pulled into orbit around it, and gradually over a long period of time it would spiral closer and closer to it. Gravitational forces would pull gas from the star until finally the black hole would be encircled by a gigantic ring—an accretion ring, similar to the ring around Saturn but much larger.

If this black hole had a mass of, say, a billion suns, and was at the center of the solar system, its surface would intersect Saturn's orbit, and its accretion ring would be a hundred times farther out than Pluto. Still, if it was in a quasar in the outer reaches of the universe, we wouldn't be able to see it, even with the most powerful telescopes.

Gas and stellar debris in the outer sections of the accretion ring gradually spiral closer and closer to the black hole. Heat is generated as particles within the disk begin to collide. The temperature reaches hundreds of thousands of degrees, then millions just before it is pulled through the surface of the black hole. The temperature is so high just outside the black hole that highly

Schematic of a black hole showing accretion disk and jets.

energetic x rays and gamma rays are generated that flood into the space around the black hole at an incredible rate. No other known source can produce more energy.

Thus, quasars contain massive black holes with large accretion rings around them. The matter from the accretion disk powers the quasar, and we see it as an incredibly energetic source. Even though this model is still speculative, it has been successful and it has been extended to other energetic objects in the universe. Closer to us, but still far away, are active galaxies. They are strong sources of radio waves, and in many cases their cores appear to be in a state of upheaval. Many have oppositely directed jets spurting hot ionized gas into the region around them. They also appear to contain black holes. The mass of the black hole in this case, however, would not be as large—hundreds of millions of solar masses, instead of billions—but the mechanism for energy generation is the same.

Jets are occasionally seen in quasars, but they are more common in active galaxies. If the black hole model applies to these objects it must explain the jets—and indeed it does. According to calculations some of the matter near the center of the accretion disk manages to avoid being pulled into the black hole. Instead, it is squeezed to a high pressure, then forced out through two tiny orifices. It emerges with a tremendous velocity, like the hot flame of a welder's torch, generating two immense regions of radiation.

But all engines are vulnerable—even the ones driving quasars and active galaxies. They depend on fuel, and when the fuel runs out they die. What happens when the fuel around a massive black hole is depleted? The black hole will still be there, but it will no longer give off energy; it will be a docile black hole. Still, it will have an intense gravitational field and if stars, gas, or stellar debris come close to it they will fall into orbit around it.

Stars orbiting black holes are of particular interest to astronomers. Their orbital speed depends on how far they are from the black hole, but those that orbit close to it travel at tremendous speeds. To the astronomer they are the signature of a black hole.

JOHN KORMENDY

John Kormendy, who is now at the Institute for Astronomy in Hawaii, was familiar with the theoretical ideas about quasars and active galaxies when he began working on the structure of galaxies. He was troubled, however, by the fact that there was no real observational proof that massive black holes existed in either quasars or active galaxies. The energy output appeared to be explained with the model, but this didn't mean it was necessarily correct. He was convinced that a more direct proof was needed.

John Kormendy.

Born in Graz, Austria, Kormendy came to Welland, a small town near Niagara Falls, Ontario, when he was three. He attended high school there, then went to the University of Toronto where he majored in physics and math, specializing in astronomy in his third and fourth years. It was during this time that he first became interested in galaxies. A visiting astronomer, Kevin Pendergast, gave a series of lectures on galaxies one summer. "He talked about the dynamics of galaxies, and how to model them on the computer. He did amazing things," said Kormendy. "His computer models looked exactly like spiral galaxies. That caught my attention . . . it got me interested."

When Kormendy graduated he moved to the California Institute of Technology, hoping to follow up on his interest in galaxies. "Caltech was a superb place. Almost everyone was an inspiration," he said. "The attitude toward science was wonderful." His thesis advisor was Wallace Sargent, an expert on galaxies, quasars, and the cosmic background radiation, who had come to the California Institute of Technology from England in the early 1960s. Kormendy said that one of the most valuable things he learned from Sargent was the importance of selecting good astronomy problems. "I remember him telling me early on about what was worth working on, and what wasn't. And the importance of thinking about why we study the things we study—that stayed with me."

Kormendy also spent a lot of time discussing galactic astronomy with the well-known cosmologist Allan Sandage, who was at the Mount Wilson and Palomar Observatories (now the Observatories of the Carnegie Institution of Washington) . "It proved quite easy to talk to Sandage because we were both interested in galaxies. He didn't like to take students on, so he wouldn't take me on as a thesis student, but he was perfectly willing to talk about science, and that is what it takes to learn science."

Another valuable experience for Kormendy while he was at the California Institute of Technology was a reading course he took from Peter Goldreich. He read everything he could on galactic structure and had a meeting with Goldreich each week to discuss what he had learned. "That course turned me into an observer with a decent theoretical background. So now when I'm doing something on galaxies, I'm not completely in the dark about the theoretical background."

When he graduated, Kormendy went to Berkeley for a two-year postdoctoral fellowship, then to Kitt Peak for a year and a half. While he was at Kitt Peak he got a permanent position at the Dominion Astrophysical Observatory in Victoria, Canada. It was during his stay at Victoria that he began using the Canada–France–Hawaii telescope on Mauna Kea for the first time. He came to the Institute for Astronomy in 1990.

Kormendy began working on galaxies during his Ph.D. thesis, and he has been working on them ever since. It soon became clear to him that the least understood and most intriguing part of a galaxy is its center. "A beautifully detailed theoretical edifice existed at the time that explained galactic nuclear activity based on the guess that there was a black hole at the center. But nobody had detected one directly. So there was this burning question: Is it possible to detect these black holes directly? This problem meshed perfectly with my developing interest in cores. I had been working on them long enough that I felt that I could recognize something peculiar or different from the ordinary."

There was, however, a problem in studying the region close to the center, namely spatial resolution. At the best of times the telescope in Victoria only had a resolution of only 2 seconds of arc. This wasn't acceptable. But now Kormendy had access to the Canada–France–Hawaii telescope on Mauna Kea, and on good nights it had a resolution of 1/2 arc second.

In addition to high resolution, however, good instrumentation was also needed. Photographic plates were not suitable for studying galactic cores, but when CCDs came along it became possible to make the required measurements. Kormendy's first serious work on the cores therefore came with the introduction of CCDs. He didn't start with active galaxies, however. Most of them were too far away, and by now astronomers were beginning to believe that all galaxies had black holes in them. The only difference between an active galaxy and an inactive one was a fuel source. Kormendy therefore measured the brightness distributions across the cores of several dozen nearby galaxies. If a galaxy did, indeed, have a massive black hole at its center, it would attract stars to it—large numbers of stars—and they would fall into orbit close to it. The core would therefore be affected.

Later, as better spectrographs became available, Kormendy began looking at the velocities of the stars close to the core. If there was, indeed, a black hole here, the stars in close orbit around it would be traveling at tremendous speeds, and you could use these velocities to determine the mass of the material inside their orbits.

Part of the impetus for studying cores and massive black holes had come from two papers that had been published by Peter Young, Wallace Sargent, and several colleagues in 1978. Most of the work on black holes in galaxies before then had been very speculative. At the time there was a theoretical model for the core of a galaxy, referred to as the isothermal model. As the name implies, the model assumes that the velocity of the stars is the same at every radius and in every direction.

Young and his colleagues used a CCD to look at the giant elliptical galaxy called M87, a strong radio source. It was a good candidate for a massive black hole in that it was active and had a jet. They found that the only way M87 could be made to correspond to the isothermal model was if there was a huge black hole at the center. The mass, they determined, had to be five billion suns.

It appeared to be a tremendous breakthrough. Proof had been found at last that there was, indeed, a black hole at the center of an active galaxy. But soon the discovery was enmeshed in controversy. By the mid-1980s astronomers, Kormendy among them, had shown that all galaxies had a brightness profile similar to M87, and therefore none of them fit the isothermal model. About the same time theoreticians also discovered that the isothermal model could not be applied to giant elliptical galaxies (galaxies with elliptical shapes), and M87 was an elliptical galaxy.

Did this mean M87 didn't have a black hole in its core? Not

necessarily, but it did tell Kormendy that the case hadn't been proved. Kormendy believes that there is, in fact, probably a black hole there, but that it is difficult to extract the information needed to prove it. "The central part of the galaxy is clobbered with light from the active nucleus; in other words, with light that has nothing to do with the black hole," he said. "It confuses the issue."

In his effort to prove, or disprove, the existence of massive black holes, Kormendy has concentrated on inactive, nearby galaxies. He has studied the cores of several galaxies, including the Andromeda galaxy (M31), its elliptical companion M32, NGC 3115 (NGC stands for New General Catalogue), NGC 4594, NGC 3377, and others. In each case he measured the brightness across the center, noting in particular how it changed at the center. He also measured the velocities of the stars near the center. The two best candidates for massive black holes at the present time, according to Kormendy, are the Andromeda galaxy and NGC 3115. Both are inactive galaxies. "They're boring as far as activity goes," said Kormendy. "But that isn't an accident. If there was nuclear activity in the center it would mask the signature of the black hole. The Andromeda galaxy is a particularly good candidate because it is so close that we can get good spatial resolution. We can put strong constraints on the mass at the center."

If you look at the brightness of the core of the Andromeda galaxy you see that it increases gradually as you approach the center, then flattens off at the center. The increase near the center is caused by a rapidly rotating star cluster in the region. Both the rapid rotation and the rapid increase in random velocities toward the center are indicative of a large mass.

Is Kormendy convinced that there is a black hole in these objects? He won't commit himself. "We're convinced that we have shown that there has to be a central object whose size is less than a few tenths of an arc second, and whose mass is about a ten million solar masses. That's not proof that you have a black hole; it's only proof that you have a central object that is dark and has a very large mass."

As we saw earlier, a black hole with a mass of 1 billion suns at the center of the solar system would extend out to Saturn's orbit (approximately 1 billion miles). Kormendy has shown that there are 1 billion solar masses inside a radius of about 3 to 4 light-years, which is about 20 trillion miles. Space considerations therefore tell us it is possible that the mass is not a black hole.

What are the alternatives? It could be a cluster of white dwarf stars or neutron stars; both are dense objects left when a massive star collapses. It is even possible it could be a cluster of small black holes, rather than a single large one. Kormendy believes, however, that there are serious arguments against all of these alternatives. "The cluster of remnants [white dwarfs or neutron stars] would have to have considerably more mass [per unit volume] than the stars in the region," said Kormendy. "So 90 percent of the mass in the region would have already evolved into remnants. But the stuff that is visible there looks normal. I don't think that such a large fraction of the stellar population could evolve into stellar remnants without polluting what's left."

How could you prove beyond a doubt that it's a black hole? According to Kormendy you would have to measure velocities extremely close to the center—so close that they are relativistic (at least 1/2 the speed of light). The radius at which this occurs, however, is less than 1/10,000 of an arc second—far less than anything we can now measure.

The next step in the search is obviously to increase the resolution. The Keck telescope may help, and Kormendy will have access to it. Furthermore, the Hubble Space Telescope, when it becomes fully operational, may also help, and Kormendy is part of a team that has received time on the Hubble Space Telescope to look for a black hole. But neither of these will take him anywhere close to the resolution he needs.

As his work has progressed, Kormendy has begun to shift some of his effort into statistics. He is interested in answering questions such as: What fraction of galaxies have black holes? What is the distribution of black hole masses relative to galaxy masses? About half of his time is now spent on this part of the project.

The Andromeda galaxy, which may have a black hole in its core. (Courtesy National Optical Astronomy Observatories)

Close-up of the core of the Andromeda galaxy at the lower left. (Courtesy Lick Observatory)

QUASARS AND BLACK HOLES

The picture we have presented of a quasar being powered by the gas in an accretion ring falling into a black hole at its center is an oversimplification. Recent developments show that things may be more complicated than this.

Cosmologists now believe that some and perhaps all quasars are colliding galaxies. According to a current model, each of the galaxies in the collision has a gigantic black hole at its core, and when they come together gas from one of the galaxies feeds the black hole in the other, and energy is released. We see the brightening as a quasar. But quasars are extremely distant objects. Some are so far away in space and time that they may have appeared when the universe was less than 1 billion years old (it is now about 15 billion years old). This means that the gas-rich galaxies that created them had to exist even earlier, and this presents a problem: there does not appear to be enough time for galaxies to form.

Another problem relates to how many galaxies become quasars. At the present time only about one quasar exists for every 100,000 galaxies. But astronomers have shown that 10 billion years ago, when the universe was young, there was about one quasar per 100 galaxies. This appears to indicate that only about 1 percent of galaxies become quasars. There is, however, a way around this. The quasar phenomenon may be a relatively short-lived phase of a galaxy's life. If it lasts for only about 50 million years, for example, it can be shown that all galaxies may go through the quasar phase.

Dave Sanders of the Institute for Astronomy in Hawaii has been working on these and other problems related to quasars for several years. His most recent interest is the molecular gas content of quasars. He and Nick Scoville of the California Institute of Technology have shown that the first quasar discovered, 3C-48, has considerable gas in it. If all quasars are gas-rich, he feels he may be able to prove that the quasar phenomenon in general is related to gas-rich galaxies in collision. He is concentrating on the

Dave Sanders.

submillimeter region of the electromagnetic spectrum, and is therefore using the James Clerk Maxwell and the Caltech submillimeter telescopes on Mauna Kea.

Sanders was born in Washington, D.C., and raised near Mount Vernon, Virginia. Science was his favorite subject in elementary school, and when he went to high school he took as much physics and math as he could. "I decided that the real test of whether you should be a scientist was whether you could make it in physics." When he went to college it seemed that physics was the field in which most of the fundamental discoveries were being made. "I guess I was primed for astronomy in that I was always thinking of the big picture," he said. "Biology and chemistry were interesting, but physics seemed more fundamental, and astronomy was just a natural extension of physics to cosmological scales."

Sanders went to the University of Virginia for his bachelor's degree, then to Cornell for a Ph.D. He started out in solid state physics, but his studies were interrupted by the Vietnam War, and when he returned he switched to astronomy. His interest in astronomy was spurred by courses he took from Carl Sagan and Frank Drake.

After he obtained his Ph.D., Sanders went to Caltech as a Research Fellow for five years. He came to the Institute for Astronomy in 1989.

Most of Sanders's early research was in millimeter astronomy (wavelengths of approximately 1 millimeter). During the 1970s and early 1980s the United States led the world in millimeter astronomy, with telescopes in Arizona, New Jersey, Massachusetts, and Texas. Gradually, however, the United States shifted to submillimeter astronomy, and most of millimeter astronomy shifted to Europe and Japan. Sanders left millimeter astronomy and went into infrared astronomy for a while, then gradually shifted to submillimeter astronomy. "It's tough working in the submillimeter region," he said. "You're always fighting opacity of the atmosphere." (The atmosphere can camouflage the results you are looking for.)

Once Sanders was at the Institute for Astronomy he had access to the only two submillimeter telescopes in the world—the James Clerk Maxwell and Caltech telescopes. I asked him why the submillimeter region of the spectrum is so useful. He said that for many astronomical objects much of the luminosity comes out in the submillimeter. This is particularly true of the type of quasar called an infrared-loud quasar. They appear to be dust-shrouded quasars in which the optical, UV, and x-ray emission is absorbed by surrounding dust and re-radiated at infrared and submilli-

meter wavelengths; some people refer to them as "cocoon quasars." The luminosity that comes out in the submillimeter region
also includes emission lines from gas molecules, which allows you
to get a measure of the molecular gas content of the quasar. In the
case of infrared-loud quasars, they appear to contain enormous
amounts of molecular gas.

Although many astronomers working on quasars now accept
the idea that quasars are colliding galaxies, the evidence appears
to be strong only for the nearest ones. Sanders believes, however,
that this picture applies to all quasars, even the most distant ones.
"For low-redshift quasars it is quite clear that a merger of two
massive gas-rich spiral galaxies is involved," said Sanders.
"Maybe at high redshifts [large distances] it's the merger of two or
more galaxies. We can't say for sure. But I believe a collision and
mergers are involved."

The idea that galactic collisions and mergers are involved at
large distances fits in well with other ideas about the universe. As
you look out farther you are looking at a younger universe, and
therefore (assuming the big bang is valid) a smaller universe. The
galaxies in it would therefore be closer together, and collisions
would be more frequent.

A recent discovery appears to support Sanders's ideas.
Michael Rowan-Robinson and several colleagues at Queen Mary
College in England, have discovered an object of high infrared
luminosity, out past the nearest quasars, that appears to be a very
young galaxy or possibly a pair of merging galaxies. It has twice
as much molecular gas in it as the most gas-rich nearby quasar.

Very young galaxies that are composed mostly of gas are
referred to as protogalaxies. In a protogalaxy most of the stars are
just forming, and there is still a large amount primordial hydrogen
gas present. Sanders believes that the Rowan-Robinson object, as
it is called, is a merger of two gas-rich galaxies, but it is older than
the true protogalaxy. "If gas-rich galaxies in collision actually give
rise to quasars, the galaxies that give rise to the most distant
quasars, which are seen back as far as 1 billion years after the big
bang, must be protogalaxies," he said.

BLACK HOLE IN OUR GALAXY?

If there are black holes in the centers of other galaxies, it only stands to reason that our own galaxy has a black hole in it. Is there any evidence for this? Indeed there is. This is particularly good news in that the core of our galaxy is much closer than the cores of other galaxies, so it is much easier to study.

It is well-known that the center of our galaxy is in the direction of the constellation Sagittarius. With a good pair of binoculars you can easily see the rich variety of celestial objects—stars, nebulae—in this region on any summer evening. What you are seeing, however, is only a tiny fraction of what is actually there. Clouds of gas and dust that do not allow visible light to penetrate them cut off our view of most of the material in this region. We do know, however, that the center is about 25,000 light-years away. The core of the Andromeda galaxy, by contrast, is 2 million light-years away.

Although we cannot see the visible emission from the core of our galaxy—the part our eyes are sensitive to—we can detect radio waves and infrared radiation from it. Furthermore, x rays and gamma rays also penetrate the dust clouds that block our view. By studying these radiations we have been able to get a reliable and relatively detailed picture of this region.

Almost exactly at the center of our galaxy is a strong radio source known as Sgr A*, an exceedingly small source—only about 1/1000 of an arc second across. This is less than a quarter of the size of the solar system. Radio astronomers have given us a lot of other information about the core of our galaxy, but the real breakthrough has been the use of infrared. In 1967 Gerry Neugebauer of the California Institute of Technology along with graduate student Eric Becklin turned an infrared detector toward the center of our galaxy and discovered to their surprise that it was an exceedingly strong source. The radiation peaked in the direction of the source Sgr A*. They mapped the entire region, and in the process made a number of important discoveries.

All early infrared measurements suffered, however, from a

A section of the sky close to the galactic center. (Courtesy Lick Observatory)

serious drawback. Our atmosphere absorbs strongly in the infrared. Becklin therefore teamed up with Ian Gatley, a Caltech graduate student, to use the Kuiper Airbourne Observatory (KAO)—a high-flying C-141 with a 36-inch telescope. They discovered that the center of our galaxy is surrounded by a doughnutlike ring of gas and dust about 12 light-years in diameter. The center of this ring had been scoured of celestial debris, yet there were stars here.

Gatley went to Hawaii in 1979 where he used the United Kingdom infrared telescope (UKIRT) to follow up on the discovery. He concentrated on the region inside the doughnut, finding that it wasn't completely scoured after all. Clouds of shock-heated hydrogen were present; in particular, there was a ring of hydrogen gas and dust orbiting very close to Sgr A* with a velocity of approximately 100 kilometers per second.

In 1981 Don Hall, who was then at Kitt Peak (presently the director of the Institute for Astronomy), and several colleagues discovered a helium line in the spectrum of the material in the central region, indicating that gas was streaming out of (or toward) the center. Tom Geballe, who is presently Head of Operations at UKIRT followed up on Hall's discovery. He has worked for ten years now, trying to determine what is going on in this region.

Like many astronomers, Geballe started out in physics. Almost all of his education, in fact, including his Ph.D., was in physics. He started on a bachelor's degree at the University of Washington in Seattle, where his father was a professor of physics. At the end of his second year his father left on a sabbatical for Amsterdam, and Geballe went with him. When he came back he transferred to the University of California at Berkeley where he completed his bachelor's degree in 1967. He stayed on at Berkeley for his Ph.D.

"I had very little interest in astronomy during my undergraduate years," said Geballe. "There were a lot of smart students at Berkeley and I didn't feel I could compete with them; in addition I was beginning to have second thoughts about physics. Fortunately I had a friend I played football with who was a student

Tom Geballe.

of Charles Townes. He asked me to come to one of Townes's seminars [which was oriented toward astronomy]. I needed a research topic so I went, and without really understanding what the talk was about I found it to be tremendously exciting. I decided to work for Townes."

Townes had received the Nobel prize for his discoveries in quantum electronics, and now wanted to use his knowledge of this area to study the universe. Several of his students, Geballe included, built novel spectrometers for studying the galactic center and other regions; Geballe built an infrared interferometer. He received his Ph.D. in 1974 and stayed at Berkeley for another year as a postdoctoral student. By this time he had a considerable interest in astronomy, but was basically still in physics.

"After Berkeley I decided to give physics one more chance so I took a postdoctoral position in Holland in a physics lab doing infrared spectroscopy," said Geballe. In Holland, however, he met a number of astronomers, and decided once and for all to switch to astronomy. So before returning to the United States he applied to the Carnegie Institution for a fellowship at Hale Observatories and was accepted. He worked there for four years, with much of his research directed at the core of our galaxy.

While a Carnegie Fellow he got a telephone call from one of the astronomers he had met in Holland telling him that Holland had purchased a 15 percent share of a telescope in Hawaii and they would need people. Would he be interested in a position? He accepted.

At UKIRT Geballe continued his work on the galactic center. He and several colleagues confirmed Hall's discovery, and over several years mapped out the high velocity gas in the central light-year of the galaxy. They showed that the wind originated from a source called IRS16, which is about 2 arc seconds away from Sgr A*. The speed of this wind is 700 kilometers per second (1.5 million miles per hour).

I asked Geballe if he was convinced there is a black hole at the center. "I'm not sure," he said. "There are a large number of objects near the center. Whether these objects are normal hot stars, or a special kind of hot star that has lost its outer atmosphere, or miniature black holes surrounded by gas, I don't know. I don't think anybody is certain at this point."

Assisting Geballe in obtaining and reducing the data is Kevin Krisciunas. Krisciunas said, with a chuckle, that one of his main jobs was to prod everyone on the project. "So many people take data, reduce it and never trouble themselves to publish it. It isn't just a matter of getting data and dashing off a note to *Nature*. Most

Kevin Krisciunas.

of what you do is hard work—organizing it, writing it up clearly. The agony of getting it through the referees and into print is all part of it."

Krisciunas was born and raised in the suburbs of Chicago. In the early 1960s he decided he wanted to be an astronaut. By junior high school, however, he had decided to become an astronomer. He built a backyard observatory and took every math and science course he could in high school. He got a bachelor's degree in physics and astronomy from the University of Illinois, then went to the University of California at Santa Cruz for a Ph.D.; while there he used some of the telescopes at Lick Observatory. Without finishing his Ph.D. he took a job with NASA on the Kuiper Air-bourne Observatory. It was here that he met Ian Gatley, who was taking measurements on the galactic core. Through Gatley, Krisciunas eventually got a job at UKIRT (later called the Joint Astronomy Centre).

Krisciunas has been involved along with Geballe, Gatley, and others in several different projects related to the core. In one they mapped the infrared helium and hydrogen lines in the central region, and in another they checked the ionization of the material at the center.

A number of other astronomers at Mauna Kea have also made important discoveries in this region. A supergiant star called IRS7, which is about 6 arc seconds from Sgr A*, was shown to have a long cometlike tail of gas streaming from it. A number of similar streamers were found by Gene Serabyn and John Lacy using the IRTF. They found streamers of ionized neon gas in elliptical and circular orbits in this region. Under the assumption that the motion of these streamers was the result of gravity, they were able to calculate that there was a mass of 3 to 4 million suns inside of them.

It is quite possible, however, that magnetic fields play an important role in this region, and if so they would affect these clouds. In 1984, in fact, Mark Morris of UCLA and Farhad Yusef-Zahed and Don Chance of Columbia University discovered three enormous parallel arcs of gas approximately 10 to 20 light-years thick and 150 light-years long projecting out from the center. They are now believed to be high-speed particles trapped by extremely strong magnetic fields.

Magnetic fields, however, don't influence stars, and there are many stars in the clear region near the center. The average distance between the stars here is approximately 1/300 the distance between stars in our region of space. Kris Sellgren, Martina McGinn, Eric Becklin, and Don Hall, using IRTF and UKIRT, were able to measure the velocity of these stars. They found that the mass inside the stars was approximately 5 million suns.

In conclusion we now know that there is a rotating, lumpy

ring of gas and dust around the center of our galaxy. The region inside it is relatively clear, but there are rings of gas here, and many stars. And finally there may be a black hole of several million solar masses at the center. Many astronomers throughout the world have contributed to our knowledge of this region, but major contributions have been made by astronomers working at Mauna Kea.

CHAPTER 10

Surveying the Universe

Understanding the core of our galaxy and the cores of other galaxies—finding out whether they contain massive black holes—is an important part of the research on Mauna Kea, but it is only one of many projects. Another problem that is being investigated is the large-scale structure of the universe. Astronomers have known for years that the universe is made up of galaxies and that these galaxies tend to cluster. More recently they have discovered that clusters also prefer the company of other clusters. Astronomers call these groups superclusters.

What are the kinematics of these giants of the universe? How do they react to one another? How do they move? One of the first great insights into the mysterious workings of the universe came from Edwin Hubble of Mt. Wilson in California. Hubble showed in his landmark paper in 1929 that galaxies are all running away from one another; in short, the universe is expanding. A measure of how fast it is expanding came from Hubble's now-famous plot—a plot of distances to the galaxies versus their velocity of recession. The slope of the line in this plot is called H in honor of Hubble.

Recessional velocities can be obtained from what is called the spectrum of a star. As we saw earlier, if you pass a beam of light through a prism, the prism breaks it up into the colors of the rainbow. The tiny particles—photons—that make up the light beam vibrate as they move through space, with photons of different "colors" vibrating at different frequencies (number of vibrations per second). The spectrum of a star, or galaxy, is obtained when the light from it is passed through an instrument containing a prism, called a spectroscope. The photons are spread out into their component frequencies, giving a series of dark lines. These dark lines are stellar fingerprints; an incredible amount of information about the star, or galaxy, is contained within them.

A crucial piece of this information was discovered in 1848 by Christian Doppler of Austria. Doppler showed that when a sound wave is approaching, its frequency increases. Similarly when it recedes, its frequency decreases. A few years later Armand Fizeau of France showed that the frequency of a light source changes in the same way. If a star is receding from us, its spectral lines will be shifted toward the red end of the spectrum (called a redshift), and if it is approaching, they will be shifted toward the blue end.

Redshifts played an important role in Hubble's discovery. He showed that literally all galaxies exhibit redshifts, and the more distant the galaxy, the greater the redshift. The redshifts of galaxies are easily obtained; they can be read directly from the spectra. Distances, however, are a different story. Hubble had to resort to a crude stepladder to obtain the distances he needed. He determined the distance to the nearest galaxies by observing variable stars, called cepheids, in them. Their brightness changes periodically over several days, and a relationship called the period–luminosity relationship gives a good estimate of a cepheid's distance, if its luminosity or brightness is known.

Hubble wasn't able to see cepheids in more distant galaxies so he used the brightest stars in them, assuming they were all about the same brightness. In even more distant galaxies he used the brightness of the galaxy itself; to a good approximation galaxies are all about the same size. In this way he set up a "cosmic ladder," which allowed him to determine the distance to many of the galaxies in our region of space.

He made a plot of distance versus redshift and found that the relationship gave a straight line. There was some scattering of the points around the line but this was assumed to be due to the inaccuracy in determining distances. The slope of the line gave a constant (later called H), and its reciprocal gave the age of the universe. Hubble was excited by the results, but when he used the slope of the line to determine the age of the universe, he was slightly embarrassed. It gave an age of 2 billion years, and geologists had already shown that it was at least twice that old. Hubble soon discovered that his estimate was, indeed, flawed. The period–luminosity relation he used was shown to be incorrect; it did not take into consideration the dust and gas—the interstellar medium—throughout our galaxy. When corrections were made, the age of the universe doubled. Within a few years other corrections were applied, and the estimate for the age of the universe increased further.

Hubble and his assistant Milton Humason worked on the project until the early 1950s. By then Humason was getting old and Hubble took on a new assistant, Allan Sandage, a young graduate of the new astronomy program at the California Institute of Technology. The association, unfortunately, was short-lived. Hubble died soon after Sandage began working for him, and the job of continuing the program fell on Sandage's shoulders.

Sandage was meticulous, and not easily satisfied. He took a long careful look at Hubble's cosmic ladder, saw how shaky it was, and decided to start over. For years he worked on a new ladder. Eventually he teamed up with Gustav Tammann of Switzerland and together they arrived at what they believed was the correct Hubble constant—50, corresponding to an age of 20 billion years. For several years it was the accepted age of the universe.

But just as Sandage had looked carefully at Hubble's work, other astronomers eventually began to scrutinize his work. The first to do so was Gerard de Vaucouleurs of the University of Texas. De Vaucouleurs went through Sandage's papers in detail and convinced himself that Sandage had not been as careful as he had led people to believe. De Vaucouleurs therefore built his own cosmic ladder and came up with an H of 100, corresponding to an age of 10 billion years.

Sandage was furious. He checked his calculations and observations carefully and claimed he had made no errors: H was 50.

BRENT TULLY

Brent Tully, who is now at the Institute for Astronomy, heard about the controversy when he was a student, but he was too busy working on his thesis to worry about it. Within a few years. however, he would be at the center of the controversy.

Born in Toronto, Canada, Tully moved to Vancouver when he was young. He had little interest in astronomy when he registered at the University of British Columbia. Most of his friends were going into engineering and his father was an engineer, so en-

Brent Tully.

gineering seemed like a reasonable choice. The first year curriculum was general, however, so he didn't have to make up his mind until the second year.

But when he finished first year he began to have doubts. He liked the physics course he had taken, and the physics program began to look more appealing than engineering. He finally decided to go into physics, but after four years of mechanics, optics, electricity, and magnetism, he was disappointed. Everything he had studied had been discovered many years earlier. "I don't think I learned anything that had happened after 1932," said Tully. So he switched again, this time into astronomy.

He went to the University of Maryland for his Ph.D. "When I switched to astronomy it seemed that everything we did was related to what was happening today. After one week in a class the instructor would say, 'I can't tell you any more because we're up to the current state of the art.'" He laughed. "I thrived on that. I knew I was in the right place; this was the place things were happening."

Tully's advisor at the University of Maryland was Tom Matthews, who had earlier worked with Sandage and Schmidt on quasars and radio galaxies. His thesis was on the kinematics and structure of M51, a galaxy in the constellation Canes Venatici that was known as the Whirlpool galaxy because of its resemblance to a whirlpool. M51 had a tidal bulge on one side and Tully's job was to explain it in terms of gravitational interactions with a nearby galaxy. "I spent a long time on my thesis, but my attitude was that it was an opportunity to do something worthwhile to get me 'jump started' in the field. I was confident it would be a successful thesis."

During graduate school Tully met Richard Fisher. Fisher had come to the University of Maryland at about the same time as Tully, and they would graduate in the same year. Just before graduation they began talking about what they would work on after graduation. They agreed to collaborate on a redshift survey of faint, gas-rich dwarf galaxies. During this collaboration a problem came to Tully's attention. He noticed that two members of the

Local Group of galaxies (the cluster of about 25 that includes our galaxy, the Milky Way), Andromeda and Triangulum, had quite different spin rates. Andromeda was rotating rapidly at about 250 km/sec, whereas the Triangulum galaxy rotated at only 100 km/sec. Andromeda was larger, and therefore more massive than the Triangulum galaxy. "Because these galaxies are so familiar, we know that one is a giant, and the other a pip-squeak," said Tully. "I asked myself: What if we didn't live right next door to them? What if they were just part of a field of distant galaxies? Would you be able to determine that one of them was a giant, and the other small? I realized that their rotational motion might tell us."

Tully decided to look into the possibility that more massive galaxies rotated faster than less massive ones. Was there, in fact, a direct relation between rotational rate and mass? This made sense in that the faster a galaxy rotated, the more mass it needed to supply the requisite centripetal (spin) force. Furthermore, there was a relatively easy way to determine how fast a galaxy was rotating. The key was the 21-centimeter (cm) hydrogen radiation that galaxies emitted. It had been shown in the late 1940s that spinning hydrogen atoms occasionally change their direction of spin via a "spin flip," and in the process they emit radiation with a wavelength of 21 cm. All one had to do was examine the 21-cm radiation line from the galaxy. Because of the Doppler shift, the radiation from one side of the galaxy (the side receding from us) would be redshifted, and radiation from the other side (approaching) would be blueshifted. The overall result would be a broadening of the line. A measure of the breadth of the 21-cm line from the galaxy would therefore give a measure of its rotational speed, and thus its mass. Then, assuming it was composed of an average distribution of stars, you could determine its true brightness. This coupled with its apparent brightness would give its distance.

The only thing needed to complete this picture was a calibration of the scale, and this could be done using nearby galaxies such as Andromeda or the Triangulum galaxy. If distances could be obtained this way, astronomers would have a powerful new tool.

Twenty-one-centimeter lines could easily be obtained using

radio telescopes. But how could the method be tested? The best way would be to use a cluster of galaxies; all of the galaxies in a cluster are roughly the same distance away. The brighter ones would therefore be more massive, and you could test to see if they had broader 21-cm lines. "What I was really pleased with in the way things developed was that the concept came before we went to the telescope. We went to the telescope looking specifically for

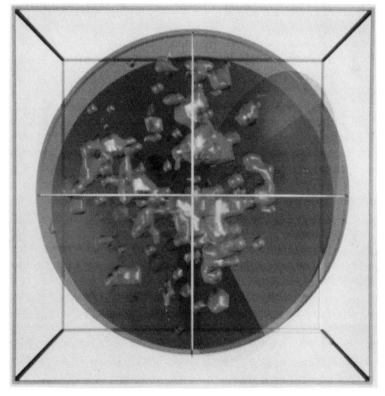

Distribution of rich clusters of galaxies. This is a region almost 2 billion light-years across. It is the Pisces-Cetus Supercluster. (Plotted by Brent Tully.) (Courtesy Brent Tully)

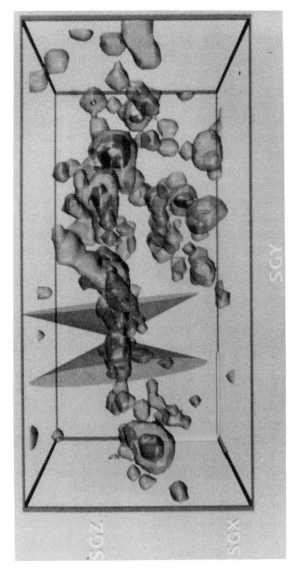

The distribution of nearby galaxies. The box is about 200 million light-years across. (Plotted by Brent Tully.) (Courtesy Brent Tully)

the relationship. It wasn't a situation where we stumbled on it accidentally," said Tully.

Tully talked to Fisher about the idea and, as it turned out, Fisher was now at the National Radio Astronomical Observatory (NRAO) in Virginia. He would be able to obtain the 21-cm lines. Tully began looking for test objects. "Fisher observed and I identified targets," he said. "I spent an awful lot of time going through the Palomar Sky Survey charts, searching for candidates to look at. I managed by the end of the time to have actually looked at the entire sky in considerable detail. I don't think many people can say that. Anyway, I came up with a list of candidates and Fisher observed them."

They decided to test the idea on the Virgo Cluster. There are about 2500 galaxies in this cluster, and they are all roughly the same distance away. Tully was now on a postdoctoral in France, at the Marseilles Observatory. He had made an extended holiday of his trip, seeing as much of the world as possible on his way to France, and when he got there, Fisher had already sent him a list of 21-cm linewidths. Plotting the data up with the appropriate corrections Tully was pleased to find that there was, indeed, a relationship. The accuracy wasn't exceedingly high—only about 20 percent—but it was clear that luminosity and mass were linked.

Tully and Fisher were excited about the breakthrough, and immediately wrote it up for publication. Strangely, though, they had a hard time getting it published. The major problem was that it gave an H quite different from what had become the accepted value. "Our H was inconsistent with Sandage's," said Tully. "But at the time de Vaucouleurs was saying, 'I don't like Sandage's methods and I don't think they're right.' So there was some controversy."

Tully and Fisher, in fact, got an H very close to de Vaucouleurs's value. Sandage heard of the result well before their paper was published, and he was not pleased. He and Tammann published a paper disputing it even before it appeared in print. They said that Tully and Fisher had not made all of the required corrections; one of their major flaws according to Sandage was that they never corrected for "Malmquist bias." This occurs because there is a tendency when dealing with a field of galaxies to select galaxies brighter than the mean. The effect is to overestimate the distance to the galaxies. The farther away the galaxy, the greater the error. Sandage, in fact, had accused de Vaucouleurs of this earlier.

One of the major difficulties was that Tully and Fisher had made no corrections for the dust obscuration in spiral galaxies. Literally all spirals have dust in them, and this tends to redden the stars. Tully was measuring blue magnitudes on his photographic plates.

Tully returned to the United States in 1975, going immediately to a job at the Institute for Astronomy in Hawaii. En route, he stopped to give a colloquium at the California Institute of Technology. Sandage was in town, and as might be expected, there was a confrontation. Marc Aaronson, a Ph.D. student who was just finishing up a thesis on the infrared properties of galaxies, was in the audience. Aaronson realized that if the magnitudes of the galaxies were measured in the infrared rather than the visible as Tully had done, the problems with dust could be overcome. Within a short time he teamed up with John Huchra, who had graduated from Caltech earlier, and Jeremy Mould, an Australian. Using a telescope at Kitt Peak they measured galaxies in both Virgo and Ursa Major, and found that the Tully–Fisher relationship was even more distinct in the infrared than in the visible. When the infrared luminosities were plotted against 21-cm line-width there was little scatter in the plot.

But Aaronson, Huchra, and Mould soon discovered that they also had a problem. The value of H they got using the galaxies of the Virgo Supercluster agreed with neither Sandage's nor de Vaucouleurs's; it was in between. After checking everything carefully they decided to publish anyway. Later, when they extended their method to more distant galaxies, well beyond the Virgo Supercluster, they were surprised to find that H was approximately 90—quite close to de Vaucouleurs's value, corresponding to an age of roughly 11 billion years.

This did not deter Sandage. He held fast with his H of 50.

CONTINUING CONTROVERSY

Tully was at the midway station, Hale Pohaku, getting ready to go to the summit, when I caught up with him. He appeared in the lobby a few minutes after I got there and shook my hand. He was tall and lanky with blond hair, and appeared to be easygoing. He excused himself briefly to get some reprints for me, and to get himself a glass of orange juice.

We went to the library overlooking the dining room. He sat down on one of the soft couches and stretched out his legs. There were large windows all around us, and a small balcony just outside. With a blue sky overhead, it looked inviting but I knew it was chilly. Just as we began to talk, another astronomer came up and asked Tully about the conditions at the summit. I had heard that they had been poor during the previous several nights, and their conversation bore that out.

Tully occasionally adjusted his glasses as he talked to me, and at times he became quite animated, demonstrating sizes and distances with his hands and fingers. Occasionally he laughed heartily.

"There has been a controversy for over 10 years now," he said. "And it perpetuated itself as long as there were no confirmations. In the last couple of years, however, we've made some real progress. People have offered a number of resolutions, and a lot of them have been well-founded. Some of them have added fine-tuning to our ability to measure distances, but I can say that the methodology that I'm using today is really not very different than it was 15 years ago. Our magnitudes had large uncertainties because they were based on photographic material or aperture photometry. People went into the infrared and that had some advantages. But we're now back in the optical using CCDs, and I think that's where we're getting the best results. CCDs have made a big difference."

Tully went on to talk about several methods that now appear to verify his results. One of these methods was introduced by Sandra Faber and Robert Jackson of Lick Observatory. The Tully–

Fisher relationship is good only for spiral galaxies; Faber and Jackson found a similar method for elliptical galaxies. Elliptical galaxies don't spin as spiral galaxies do. The stars in them travel in many directions around the core, but this also causes the spectral lines to broaden. Faber and Jackson found a relationship between the breadth of the line and the measured size of the galaxy, and it also gave the distance to the galaxy. The method has been modified in recent years and now appears to give distances to an accuracy of about 20 percent.

More recently a planetary nebula method has been discovered. Late in the life of certain stars they blow off their outer layers in what looks from a distance like a smoke ring. Calculations show that there is an upper limit to the brightness of this ring, and this has allowed astronomers to determine the distance to galaxies in which they have been observed.

Another method makes use of the uneven brightness, or mottling of galaxies. Galaxies are, of course, composed of stars, some of them giant stars, and because of this they are not uniformly bright. A measure of the mottling of galaxies has given astronomers a method for estimating their distance.

Tully is pleased that all of these methods are giving results very close to what he is getting, namely that H is approximately 90. He mentioned, however, that there were two other methods, both involving supernovae, or exploding stars. They were associated with the two basic kinds of supernovae, Type I and Type II. Type I supernovae are associated with young stars—stars in the arms of spiral galaxies. Type II are associated with older stars such as those in the core of spirals, or in ellipticals.

"Until a month or so ago both of the supernova methods gave a low H, close to Sandage's," said Tully. "But at a recent meeting in Aspen, the Type II supernovae were shown to give a high H, in agreement with ours. So there are now four good methods giving a high H, and only one arguing for a low one."

Tully is convinced that his value is correct. "If the higher value is wrong, you've got to say that four methods, all of which have nothing to do with one another, somehow have a flaw in them. And in each case the flaw ends up giving the same H." He shook his head. "I would be very surprised if that was the case."

In 1988 Tully published a paper outlining what he believed were the flaws in Sandage's method. He pointed out that our galaxy lies in what he calls the Coma-Sculptor Cloud. The Coma-Sculptor Cloud is roughly in the form of a cylinder—50 million light-years long by about 5 million light-years across. We are near one end. The ends of the cylinder point at the constellations Coma Berenices and Sculptor. Objects within this cloud are falling toward one another with an average velocity of 100 km/sec. This has a significant effect on the value of H as it severely retards the outward or Hubble expansion of the universe. Tully refers to it as the local velocity anomaly.

It has been known for years that as you go outward to more distant galaxies the value of H grows, but Sandage attributes this to Malmquist bias, and says it should be corrected for. Tully disagrees; he is sure that any corrections needed for Malmquist bias are small. In particular, he points out that if Malmquist bias is the problem, the value of H should continue to grow as you go to increasingly distant galaxies. But it doesn't. It takes a significant jump just outside the Coma-Sculptor Cloud, then levels off.

PECULIAR VELOCITIES AND THE GREAT ATTRACTOR

While there is still some controversy over H, most astronomers have moved on to the fine structure in the expansion of the universe. Are there local deviations from the smooth, uniform Hubble expansion that characterizes the entire universe? If the matter of the universe is not uniformly distributed on a large scale, local deviations would be expected because of uneven gravitational pulls. We have, in fact, already seen one example of this. The galaxies within the Coma-Sculptor Cloud pull one another around with an average speed of about 100 km/sec.

What about larger objects at greater distances? They may have an even greater effect, depending on their mass. If we are to

measure "peculiar velocities," or differences from the Hubble expansion, we will, of course, need a frame of reference. As it turns
out we have one: the cosmic background radiation. It fills the
universe, and any motions within the universe can be compared
to it. Such measurements have been made to a high degree of
accuracy recently by the satellite COBE. According to COBE our
Local Group of galaxies is traveling toward the constellation Leo
with a velocity of about 600 km/sec. This is our peculiar velocity
relative to the universe.

What is causing it? Some of it, as we just saw, is caused by the
Coma–Sculptor Cloud. A second source is the supercluster that
contains our Local Group, a supercluster called the Local, or Virgo, Supercluster. Our galaxy lies near its edge; at its center is a
particularly large cluster called the Virgo Cluster. The Virgo Cluster is about 50 million light-years from us, and is so massive that
it exerts a considerable pull on us. Astronomers have, in fact,
shown that the Coma-Sculptor Cloud is moving toward it at about
300 km/sec. But this isn't the end of the story; all of the 600 km/sec
is still not accounted for.

What is causing the additional velocity? Several years ago a
group of astronomers, who referred to themselves as the Seven
Samurai, set out to find the answer. They included Alan Dressler,
Sandra Faber, Roger Davies, Donald Lynden-Bell, David Burstein,
Roberto Terlevich, and Gary Wegner. It is not an easy task to find
a large mass in deep space that is influencing the motion of our
galaxy. The difference in velocity in the direction of this excess, as
compared with the Hubble velocity, would be tiny. Furthermore,
the redshifts of hundreds of galaxies would be needed, and an
accurate measure of their distances would be a necessity. The
Tully–Fisher relationship would give these distances. As we saw
earlier, however, Faber and Jackson had developed a similar
method for elliptical galaxies, and the Seven Samurai used it.

Over a period of several years the Seven Samurai plotted the
excess velocities of hundreds of galaxies, some far beyond the
Virgo Supercluster, and they found a huge excess in the direction

of the constellations Hydra and Centaurus. There was, in fact, a giant supercluster in this direction, called the Hydra-Centaurus Supercluster, but the excess could not be accounted for by it alone. They therefore postulated a mass beyond it called the Great Attractor; according to their calculations it was giving the Local Group a velocity of 530 km/sec.

How big is the Great Attractor? Current estimates put the galaxy count at 7500. Interestingly, though, these galaxies cover a particularly large volume; as seen from earth the Great Attractor encompasses almost a third of the southern sky. Because they are in such a large volume, it doesn't seem as if these galaxies would have enough mass. It is possible, though, that they contain considerable "dark matter." Dark or invisible matter is known to exist in literally all clusters.

Strangely, even the Great Attractor doesn't account for all of the velocity; 370 km/sec in the direction of the constellation Canis Minor is still unaccounted for. Furthermore, at the opposite end of the sky from the Great Attractor is another large supercluster called Perseus-Pisces. It is slightly farther away than the Great Attractor but should attract us with just as great a force, and there is still some confusion over whether it does or not.

The problem of peculiar velocities is intrinsically tied up with the problem of measuring H. In his attempt to obtain an accurate value of H Tully had to deal with peculiar velocities, and this eventually led to an interest in the peculiar velocities themselves. For the last few years he has, in fact, been spending most of his time on this project.

In 1990 Tully and colleagues Edward Shaya of Columbia University and Michael Pierce of the Dominion Astrophysical Observatory in Canada took a fresh approach to the problem of peculiar velocities using data from several sources, including Tully's *Nearby Galaxy Atlas* and the IRAS catalogue. They also made measurements on 300 galaxies. Contributions from the Coma-Sculptor Cloud and the Virgo Supercluster were included, and like the Seven Samurai, this group found a large attraction coming from the direction of the Great Attractor. To make their model work, however, they found they needed a number of other contributions.

Edmund Bertschinger of MIT and Avishai Dekel of the Hebrew University of Jerusalem had come up with a particularly good visual representation of the matter of the universe in the region of the Great Attractor and the Perseus-Pisces Supercluster. One of the things they discovered was that there was an unidentified "mystery peak" hidden by our galaxy, about 60 degrees from the direction of the Perseus-Pisces Supercluster. Tully and his colleagues incorporated it into their model. They also added a contribution in the direction of the Great Attractor. "I think the main component that is causing the flow in the direction of the Great Attractor is not the Great Attractor itself—it's behind it," said Tully. Pointing to the region in one of his diagrams he continued, "This region is a lot bigger than the Great Attractor, and it's at least three times farther away."

The region that Tully pointed to had actually been noticed as far back as the 1930s by Harlow Shapley of Harvard. It was a region with a high concentration of galaxies that is now known as the Shapley Concentration. Roberto Scaramella and several of his colleagues of Italy mentioned in a paper in 1989 that this region might be important.

"We get a local velocity of 600 km/sec by including the Shapley Concentration," said Tully. "But at this point our model is still pretty crude."

Tully's model agrees with that of the Seven Samurai in several respects, but it doesn't agree completely. Furthermore, another group working with Michael Rowan–Robinson of Queen Mary College in England has another model based on the infrared galaxies detected by IRAS, and it doesn't agree with either model.

Rowan–Robinson was born in Edinburgh and grew up in Sussex, England. His main interest when he was young was mathematics. "The first telescopes I looked through were large ones," he said. He received a bachelor's degree in mathematics from Cambridge and a Ph.D. in astronomy from the University of

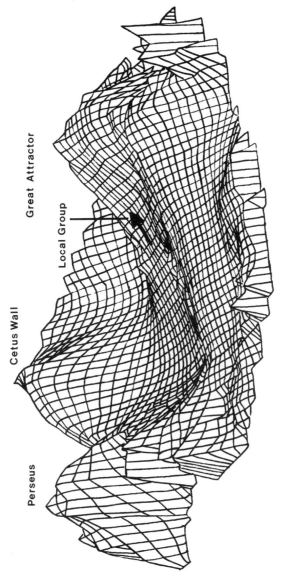

The Bertschinger–Dekel plot of the attractors in space showing the Great Attractor. Peaks correspond to regions of high attraction.

London. His switch to astronomy was partly due to the influence of the cosmologist Fred Hoyle. "I loved Hoyle's astronomy books," said Rowan–Robinson. "He was at Cambridge when I was there and I went to his lectures from time to time. He was a very charismatic figure." Rowan-Robinson worked for William McCrea on a theoretical model of quasars and radio sources while at the University of London.

For the last few years Rowan–Robinson and his group have

Michael Rowan–Robinson.

been concentrating their efforts on a large-scale mapping of the universe. "In the infrared galaxy survey we've made, we've mapped a large volume of the universe," he said. "We've made a three-dimensional map of the universe and can determine the attraction of other galaxies on our infrared galaxies. We've estimated the gravitational pulls, and when we added them all up they agree very well with the direction our galaxy is moving relative to the cosmic background radiation."

I asked him if his group had detected a Great Attractor. He shook his head, "We don't postulate a Great Attractor. We just take what we actually see and use it. I've said many times that I don't think that the Great Attractor exists. In our three-dimensional maps there is no such object. I've had some pretty good arguments with Alan Dressler [of the Seven Samurai] about this."

SANDRA FABER

Sandra Faber, one of the Seven Samurai, says she is confused by Rowan–Robinson's assertion that there is no Great Attractor. "Our data definitely show one," she said emphatically. Faber played an important early role in getting the Keck telescope project started, and she was on the Science Steering Committee for the project for several years. Now that the telescope is almost ready for research, she is eagerly waiting to use it.

Born in Boston, Faber became interested in astronomy before she went to school. In later years she spent hours looking through binoculars at the stars. But at that time she was still interested in many things; she collected rocks, leaves, and plants, and she read everything she could about science. One of the most influential books she read at this time was Fred Hoyle's *Frontiers of Astronomy.*

"I can't say I had decided to go into astronomy when I was young," said Faber. "I was a girl, and it wasn't clear that girls did things like that. I didn't come from a scientific or academic family so it was a mystery to me who did astronomy. I didn't know any

Sandra Faber. (Courtesy Lick Observatory and Don Harris)

astronomers, and as far as I could tell from the books I read, astronomical facts came from geniuses like Einstein and Hoyle. It wasn't obvious that an ordinary person, which I considered myself to be, could become an astronomer. So I didn't think too seriously about it."

When she graduated from high school she applied to Swarthmore College. Under "interest" on the college application form she wrote, "I'm interested in studying the universe." But she admitted that she wasn't sure what that entailed. She decided to major in astronomy and chemistry but soon discovered that she didn't like chemistry. Fortunately, after talking with some of her instructors, she found that physics would be of more value to her if she was interested in astronomy. She worked with several astronomers at Swarthmore, including Sarah Lee Lippincott, and Peter van de Kamp, who later became famous for discovering dark objects, possibly planets, around several nearby stars.

Faber went to Harvard for graduate work. Her thesis was on the photometric properties of elliptical galaxies. Elliptical galaxies are different from spirals such as the Andromeda galaxy, in that they do not have arms, contain almost no gas, and are generally composed of older, redder stars. As their name implies, they are elliptical, but their ellipticity can vary from very elongated to round.

Faber completed her Ph.D. in 1972 and went to Lick Observatory in California. Shortly after coming to Lick she and a graduate student, Robert Jackson, discovered the now-famous Faber-Jackson relation. I asked her about it. "It was a serendipitous discovery," she said. "When I came to Lick I was very interested in the spectra of elliptical galaxies. They showed dark absorption lines, and were, of course, just the sum of the spectra of all the stars in the galaxy. I was interested in studying these spectra with the view of understanding the stars in the galaxies better. When I arrived at Lick there was a brand-new detector that had been built by Joe Wampler and others, called the image dissector scanner. My first thought was to use it to study the spectra of these galaxies. I thought I would analyze the strengths of the absorption lines and find out more about the stars in them. In the process of taking those spectra I saw that the width of the absorption lines were different in different spectra. This, of course, was caused by the Doppler effect; in the spectra of a galaxy you are looking at the superimposed spectra of billions of stars, so you get a smearing out of the lines because the stars are going at different speeds.

"I could see this effect easily in the spectra and I thought nobody had measured it before. I said to myself: 'Why don't I just set my spectral project aside for a while and look into these results and see what I get?' So along with graduate student Robert Jackson I devised a couple of ways of measuring the smearing of the lines, and came to the conclusion that the lines were broader for brighter galaxies."

Properly calibrated, the relationship allowed them to determine the distance to a galaxy from the width of the spectral line and the diameter of the galaxy. It is, in effect, analogous to the Tully–Fisher relationship, but applies to elliptical galaxies. The discovery was made independently at about the same time as Tully and Fisher made their discovery.

Once the relationship was established, Faber wanted to make use of it, and as a result, the Seven Samurai were formed. Over the next several years they published two papers analyzing and describing galactic velocity fields, or peculiar velocities. In the first paper they concluded that all of the galaxies in our vicinity were streaming in the direction of what is now known as the Great Attractor with a velocity of about 600 km/sec. Then they took a closer look at the data and decided a better representation was an inflow from all sides to the middle of the Great Attractor. Faber now believes that both of these pictures are involved. "We now think that the Great Attractor and the entire region is moving, along with the inflow," said Faber. "And I think Brent Tully's suggestion that the Shapley Concentration is playing an important role in the motion of the whole region is probably true."

Faber, Dave Burstein, and several other astronomers have recently put together a catalogue of galactic streaming velocities. There are 3100 galaxies in it, including many measured by several other groups. Some of the measurements come from a group in the southern hemisphere.

"Our analysis from this new catalogue gives good results, and it definitely shows a Great Attractor. Regarding Rowan-Robinson's assertion that there isn't one—"; she paused. "I don't know . . . his maps based on his infrared survey don't appear to agree at all with what you see optically. Furthermore, there's another survey of infrared galaxies from IRAS headed by Marc Davis, and we get good agreement with them."

I asked Faber if she'll be using the Keck telescope to follow up on the project. "No," she said. "The Great Attractor project is not something that is particularly well-suited for the Keck telescope. Keck is special because it allows people to look to extremely faint objects, and the Great Attractor project doesn't require this." Instead she will be working on a project with several other astronomers called "Deep," which will be a search for galaxies and quasars in extremely deep space. We will turn to it in the next chapter.

CHAPTER 11

Searching for the Ends of the Universe

Looking into the sky on a starlit night you cannot help but wonder: does the universe ever come to an end? If you stop and think about it, you soon realize that an end or "boundary" would present a problem. Most people would immediately ask what was on the other side. The alternative, namely that there is no end, is just as problematic, however. It means that the universe goes on forever—to infinity. And scientists don't like infinities; most feel that in using them we are really just sweeping dust under the rug.

Fortunately, we do have a way around the problem. Nature, it turns out, limits our view of the universe. The universe may, indeed, go to infinity, but it need not worry us because we only see a finite part of it—what astronomers call our observable universe. We are limited to this region because the light from stars and galaxies travels at a finite speed (186,000 miles/sec) and therefore takes a finite time to reach us. As a result we see these objects as they were when the light beam left them. Light from a flare on the sun, for example, takes approximately 8 minutes to get to us. We therefore see the flare 8 minutes after it actually occurred. The light from some of the stars in our galaxy, on the other hand, takes thousands of years; we therefore see them as they were thousands of years ago. And beyond our galaxy, we see other nearby galaxies as they were millions of years ago. Probing further with large telescopes we see faint radio galaxies and quasars as they were billions of years ago. Finally, at 15 billion light-years (assuming our universe is 15 billion years old), we come to the end of our universe; before this it didn't exist.

What would we see if we had telescopes large enough to observe the region near the edge of our observable universe? This is where time began, and according to the big bang theory, the universe consisted of nothing more than a hot expanding gas cloud. The galaxies we see all around us today came, of course, from this cloud, so the "seeds" that produced them had to develop in the cloud. Most astronomers believe these seeds were tiny fluctuations that produced overdensities (regions of greater-than-average density). The gas surrounding an overdensity would be attracted to it, and would gradually build up until finally it broke away from the overall cloud.

The first forms in the universe, then, were nebulous balls of gas which, in time, collapsed into what we refer to as protogalaxies. Protogalaxies consisted of gas and early forms of stars—protostars. In time these protogalaxies evolved into galaxies.

This is, in theory, what we would see if we had large enough telescopes. What do we see? A mottling of the background radiation in the universe was recently detected by the satellite COBE. It may be our first glimpse of the breakup of the cloud. Closer, but still near the edge of our observable universe, we see quasars that appear to be embedded in giant "host" galaxies. We see no evidence of the host galaxies in the most distant quasars, but they can be seen in some of the nearer ones. Closer yet are active galaxies, very energetic objects that spew out thousands of times as much energy as our own galaxy. And finally nearby we see ordinary galaxies like our own.

One of the major goals of modern-day astronomy is to see as far out into the universe as possible, and that is why astronomers build large telescopes. The Keck telescope, with its 10-meter mirror, will allow them to see much farther than they've ever seen before.

As we saw earlier Sandra Faber and her colleagues will be among the first to use the Keck telescope to probe the depths of space. She discussed "Project Deep" with me. "It will have two phases," she said. "In the first phase we'll use the low-resolution spectroscope that has already come on line. In the second we'll be using a much larger spectroscope that is now being built. It will be five times faster."

The object of Project Deep is to look for the earliest forms of galaxies—galaxies in the farthest reaches of space. "We expect to be routinely surveying the universe when it was about half its present age," said Faber. "But when we look particularly deep, and find really deep objects, we will see them as they were when the universe was less than one-fifth its present age."

Several astronomers will be working with Faber on this project, including Garth Illingsworth and David Koo of Lick, Marc Davis of Berkeley, Richard Kron of Yerkes, and John Kormendy of the Institute for Astronomy.

SPECIFYING DEEP SPACE

As we look at galaxies farther and farther in space we see their spectral lines shift increasingly to the red. This means that the farther galaxies are away, the faster they are receding from us. In theory, they reach their greatest speed, namely the speed of light, at the edge of our observable universe. Astronomers specify the speed of a galaxy, or equivalently, its redshift, as a fraction of the speed of light. If a galaxy is moving away from us at, for example, half the speed of light, they say it has a redshift, or Z, of $\frac{1}{2}$. According to this, a Z of 1 would correspond to galaxies traveling at the speed of light. But Einstein's theory of relativity tells us that nothing can travel at the speed of light. It would seem, therefore, that we can't have a Z of 1. But Z's of 4 and more are known, and the reason is that the simple formula defining Z breaks down at high velocities, and must be replaced by a relativistic formula. When we speak of a redshift of 1, therefore, the object is not traveling at the speed of light. The only thing we can say is that the wavelength of all of the spectral lines in the spectrum of the object will have increased by an amount equal to the original wavelength; in other words, they will have doubled.

At a redshift of 1 it turns out that the galaxy is receding from us at a speed approximately 60 percent that of light. Since the edge of our observable universe corresponds to a recessional speed equal to the speed of light, we can easily determine how far this object is from us. Assuming the age of the universe is 15 billion years, its distance is 9 billion light-years.

The most distant object we know of today is a quasar that has a redshift of 4.8. It is receding with a speed approximately 94 percent that of light. This quasar is seen as it was roughly a billion years after the beginning of the universe. The most distant galaxy, on the other hand, has a Z of 3.8.

Quasars and radio galaxies of high Z tell us that galaxies formed before Z = 5, corresponding to a time less than a billion years after the big bang. Massive galaxies were therefore present at this time, but this doesn't necessarily mean that the bulk of galaxies formed then. Quasars and radio galaxies with high Z are relatively rare. We have detected less than 50 so far.

Also important to astronomers is the "magnitude," or brightness, of these objects. The first magnitude scale was set up by early Greek astronomers, and the scale we use today is based on it. Objects are assigned a number on this scale, with the smallest numbers designating the brightest objects. An object of magnitude 1, for example, is approximately 2½ times as bright as an object of magnitude 2. Similarly an object of magnitude 2 is 2½ times as bright as one of magnitude 3. The dimmest object we can see with the naked eye is roughly of magnitude 6. Through the largest telescopes, however, we can see to magnitude 22, and in recent years, with modern CCDs and so on, astronomers have been able to detect objects to magnitude 28.

SURVEYING DEEP SPACE

Several people are using the telescopes of Mauna Kea to probe to the farthest reaches of space. Len Cowie, of the Institute for Astronomy, has been working on a survey of deep space since he came to the Institute in 1986. His main interest at the present time, however, is not individual high-Z objects, but rather the bulk of the galaxies. He began the survey with Simon Lilly, but Lilly has since moved to the University of Toronto.

All of the work at the present time is being done with CCDs and infrared arrays. Photographic plates were used in the early days of the survey, but are very inefficient compared with CCDs, and are no longer used.

I talked to Cowie in his office at the Institute for Astronomy in Manoa. It was one of the larger offices in the building, with a desk at one end and a long blackboard covered with diagrams and equations along one wall. Cowie is tall, and has a deep voice with a slight accent. He was born in Jedburg, a small town on the border

Len Cowie.

between England and Scotland. He originally wanted to become an archaeologist, but later settled on physics. He took his undergraduate degree at the University of Elizabeth in England, then went to Harvard for his Ph.D.

He had little interest in astronomy when he first went to college. His undergraduate degree was in mathematical physics, and his Ph.D. from Harvard was in physics. His thesis, however, was astronomy-related; he worked for George Field on a project trying to determine the nature of the evolution of stellar gases. It was his first project in astronomy, and he's been working in astronomy and cosmology ever since.

Upon graduation he went to Princeton University for several years where he worked with Jerry Ostriker. They collaborated on a model of the origin of the large-scale structure of the universe. Astronomers at Harvard had discovered large spherical voids in space, with sheets of galaxies and clusters strung over their surfaces. Cowie and Ostriker showed that these voids may have been caused by supernova explosions in the early universe. The idea was interesting, but there was a problem. The largest bubbles they could generate were about 10 to 20 million light-years across; the ones that were observed by the Harvard group were 100 to 200 million light-years across.

From Princeton, Cowie went to MIT as an associate professor. He was tenured at MIT but stayed only a short time, moving on to the Space Telescope Institute in Baltimore in 1985. Until now he had been working primarily on theoretical problems related to astrophysics. But he was getting uncomfortable with some of the speculative models in the literature. "You have theoretical ideas and you want to confirm them," said Cowie. "I felt that theoretical astrophysics was decoupling from observations. There was just too much of a jump, and I wasn't happy about it, so I started doing more and more observations." But he soon found that observing time on large telescopes was not easy to get.

The director of the Space Telescope Institute when Cowie was there was Don Hall. In 1986 Hall left to become director of the Institute for Astronomy. Cowie knew that if he could get a position at the Institute for Astronomy he would have much better access to large telescopes, so he wrote to Hall, and within a few months he was in Hawaii.

"Our object when we started this program was to take extremely deep infrared images of the sky. Infrared arrays were just coming into existence, along with CCDs, and they would allow us to get the images we wanted," said Cowie. Cowie and his colleagues were interested in obtaining infrared images because distant galaxies with large redshifts were easier to detect in the infrared. They planned on starting with number counts, in other words, counts of the number of galaxies of a particular magnitude in a unit area, then following it up with spectroscopy, which would give the redshifts of the galaxies.

Most of the early observations were done using the United Kingdom infrared telescope (UKIRT), but its infrared array was relatively small, and it took about 40 to 50 hours to scan a field. In 1990, however, a 256- by 256-pixel infrared array was completed at the Institute for Astronomy, and it was used with the University of Hawaii's telescope and the Canada–France–Hawaii telescope. The time for a scan decreased considerably.

In probing to the most distant objects they found large numbers of galaxies in the field—typically about 100,000 per square degree. As they went to dimmer and dimmer magnitudes they counted the number of galaxies in each square degree and found it increased dramatically. Both visible (blue) and infrared counts were made, so they could be compared. They found that in the visible the galaxy counts increased to the faintest magnitudes, but in the infrared the curve began to flatten, or level off, at about $Z = 1$.

"A flattening was expected," said Cowie. "It has to do with the fact that the universe was smaller when it was very young, and since it was smaller you would expect to see fewer galaxies." What was strange, however, was that a flattening was expected in both the infrared and the visible. In the visible, however, large numbers of faint blue galaxies were found. Tony Tyson of Bell Labs had actually observed these blue galaxies earlier, and had

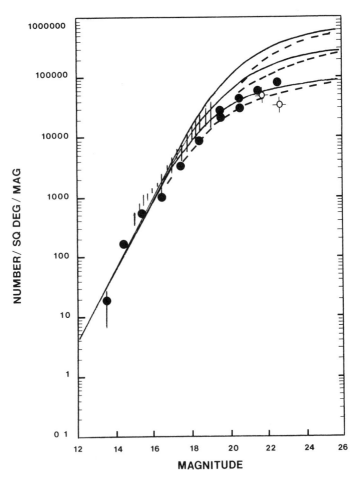

A plot of the number of galaxies versus magnitude. Note the drop-off at large magnitudes.

assumed they were extremely distant galaxies with very high redshifts.

When Cowie obtained the redshifts of some of these faint blue galaxies he was surprised to find that they weren't nearly as large as expected. They were much smaller than the dimmest red galaxies. Instead of having a redshift near 1, they had redshifts near 0.2. This meant that they were much closer than the dimmest red galaxies, and were dim only because they were small. But there were tremendous numbers of them, and astronomers knew there weren't large numbers of small galaxies in our region of space. Why, then, did such a large number of them exist over a region of deep space (i.e., in the past)?

Also, why were they blue? Highly redshifted galaxies should be red. Cowie believes that they are "starburst" galaxies, in other words, galaxies in which millions of new stars are forming. If so, they would be blue. Furthermore, if they all have a Z of 0.2 to 0.3, they are no longer with us; Cowie refers to them as a "dead" population. "For some reason, either because they were disrupted in the star-making process, or because they are now too dim, we no longer see them," said Cowie. He points out, however, that they contain a tremendous amount of mass, and may be important in relation to the dark matter problem.

The next step in the program, according to Cowie, is "to push the bulk of the galaxies through $Z = 1$"; in other words, observe and record large numbers of galaxies with Z greater than 1. So far, with the survey out to magnitude 24, very few have been seen in this region. "We're really pushing the edge now with 6- and 7-hour exposures," said Cowie. "But Keck will give a tremendous gain. A magnitude or two will make an enormous difference. Keck will really change things as far as faint galaxies are concerned." He mentioned that the new generation of spectrographs now coming on line, some of which will be used with Keck, will also be helpful.

Cowie pointed out, however, that despite the optimism, there are problems. First, if you want to understand the numbers and distribution of galaxies in deep space, you have to understand the local distributions, and in the accepted models of local distributions, galaxies are assumed to be brighter in the past. In other words, they have evolved into generally dimmer galaxies as the universe has aged. This makes sense in that the first stars to form would likely have been very massive, and therefore very bright. When these stars exploded as supernovae they would have formed smaller stars that do not, in turn, explode as supernovae. If all galaxies started with large numbers of bright stars and now contain large numbers of dim stars, they should be brighter in the past. But according to Cowie's observations this doesn't appear to be the case. Galaxies don't appear to be brighter in the past; they are actually dimmer.

How is this possible? According to Cowie, "Galaxies may have formed from pieces [smaller galaxies]. While the whole entity may now be bright, the pieces, which existed earlier, were relatively dim." He mentioned that several people are working on theoretical models trying to incorporate this idea. But there is a serious problem: spiral galaxies are difficult to produce via mergers. And the reason is that they have disks. Elliptical galaxies are relatively easy to make via mergers, but the disk and arms of spirals are not. The main reason is that the stars in spirals all go around in the same direction, in the same general plane (i.e., in the disk). The stars in ellipticals, on the other hand, orbit in many different planes, and directions. The collision of two galaxies is more likely to produce a disordered array of orbits, and therefore, an elliptical.

Cowie is sure, nevertheless, that mergers play an important role, and he is hoping to catch early galaxies in the process of merging. While there are some candidates—the Rowan-Robinson object, for example—no one has positively identified such an object.

Another way of getting around some of the problems is if the central bulge of a spiral forms considerably earlier than the disk. If you look carefully at a spiral you will see that its central bulge looks much like an elliptical galaxy. Cowie believes that mergers may produce arms later, but admits it is difficult to model.

MARKERS AND LIGHTHOUSES

Over the years Cowie has collaborated on several projects with Esther Hu, who is also at the Institute for Astronomy. She is presently working part-time with him on his deep sky survey. Her major interest, however, is very deep objects—those well out past a redshift of 1. About 40 to 50 quasars and radio galaxies have

Esther Hu.

been found in this region. She hopes that by studying them, and the regions around them, she will learn something about the earliest forms of galaxies.

I first met Hu as she was getting ready for a night's run on the Canada–France–Hawaii telescope. I was with Kevin Krisciunas, who was pulling on heavy boots and a large parka to ward off the cold of an unheated dome. Her mouth dropped when she saw his outfit. "I know," she laughed. "I've spent plenty of nights in those unheated domes. They're cold."

A few days later I saw her at the Institute for Astronomy. "What did you think of the seeing the other night?" she said in a loud, excited voice, as I walked into her office. "Wasn't it fantastic!" I agreed that it was best I had ever seen.

Her office, like many of the others at the institute, was lined with journals and books. Several photographs were pinned on a bulletin board over her computer. A large file cabinet stood in one corner. Her long straight hair was tied at the back of her head, with a few strands sticking out here and there. She talked excitedly about her research, occasionally interrupting herself, asking, "Do you understand what I'm saying?" Every so often she would jump up from her desk and grab a chart or photograph, using it to point something out to me.

Born and raised in New York City, her first contact with astronomy came in the second grade. She bought the book *Insight into Astronomy* by Leo Matersdorf. "That was my introduction to astronomy," she said, "But I was mainly excited about the book because of the introduction by Ralph Waldo Emerson." She had it memorized and proudly quoted it for me. Then, laughing, she said, "That's the 'poetic' explanation of my introduction to astronomy. I didn't really think about astronomy much after that until junior high and high school." Her first interest, she said, was physics.

"The appealing thing about physics for me was that it gave some insight into the way the universe works. I was in the generation that grew up in the space age . . . I got to see men walking on the moon. At the time I would say that most of the really interesting discoveries in physics were astrophysical ones. Almost any area you study in physics, you are studying the fundamental properties of the universe. You are studying conditions you could never reproduce in the lab ... all those exotic things like black holes and pulsars. It was a very exciting time for me."

Hu obtained a bachelor's degree from MIT and a Ph.D. from Princeton University. Her thesis director at Princeton was Ed Jenkins; she studied neutral hydrogen and its distribution in our galaxy. "I started out with our galaxy," she said, "and I've been moving out farther and farther into the universe ever since."

After graduation she worked at the Space Telescope Institute for a while. The first x-ray satellites, Uhuru and Einstein, had just been launched and she got involved in a project on x rays. Results from the satellites indicated that clusters of galaxies were emitting x rays, but astronomers had no idea how they were being generated. Later it was discovered that these clusters had hot halos, with temperatures up to 100 million degrees, associated with them. Hu became interested in how long it took these halos to cool. She found, to her surprise, that it took a time roughly equal to the age of the universe.

One of her first collaborations with Len Cowie also came at about this time. They were looking for objects with high redshifts, trying to find primeval galaxies and study their properties. On one of the photographs several of the objects appeared to be double. "It looked like over a small region of space we had double vision," she said. "About this time gravitational lensing and cosmic strings were in vogue." Gravitational lensing causes a splitting of images, resulting in double or triple images of the same object. All that is needed is a very dense object along the line of sight, and according to grand unified theories, cosmic strings were exceedingly dense. "There were three pairs in our field of view," said Hu, "and it looked like they might be caused by a cosmic string, so we published an article on them." She smiled. "It was really just a serendipitous discovery." She admitted that she's still not sure whether the phenomenon is caused by cosmic strings, but it was interesting.

Hu came to the Institute for Astronomy in March 1986, just in time to see Halley's comet. "I landed in Oahu on Friday, and had to leave the next day for an observing run on Mauna Kea," she said. "At the end of the run we were all rushing to get our equipment out of the way and get things switched over so other observers could observe Halley's comet. We were working so fast and hard to get things done that by the time we finished it was starting to get light. I rushed outside to see if I could see Halley, but it was already too light." She was disappointed, but said she saw it the next night back on Oahu, but the conditions weren't as good, and there were no large telescopes.

Hu is working on two projects at the present time. Besides working with Cowie on his deep sky surveys, she is searching the region near high-redshift quasars for evidence of early forms of galaxies. "Both of my projects are designed to find out what is happening to galaxy populations very early on," she said. "We would like to answer questions such as: when did the epic of galaxy formation begin? When and how did things really start happening?"

In the deep surveys she is interested only in what is happening to galaxy populations—the bulk of the galaxies. In her second project, however, she is looking at individual objects. She is particularly interested in the fields around quasars with redshifts greater than 4. About 30 are known.

She pointed out that there are three different ways of finding out about early galaxies by looking at these fields. "First you can use these quasars as a way of tracking down high-redshift galaxy populations," she said. "The best place to look for a galaxy is near one, since galaxies tend to cluster. So if the quasar has a particular feature, say a bright emission line, you would look for other objects with bright emission lines close to the quasar. They may have the same redshift. In this case you're using the quasar as a 'marker.'"

A second way you can use these quasars, according to Hu, is as "lighthouses." Again, you look in the immediate vicinity of the quasar, but this time you are searching for protogalaxies, or pro-

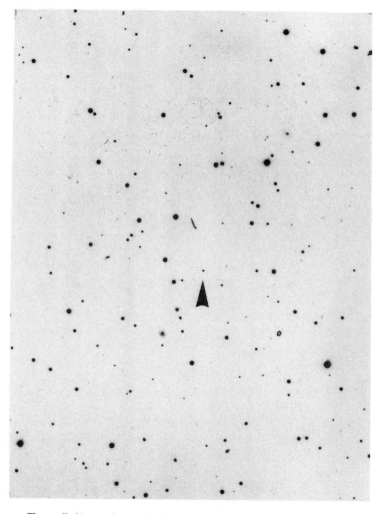

The small object at the arrowhead is a quasar. (Courtesy Hale Observatories)

togalactic material. In most cases it would be a hazy patch that would be too diffuse and dark to see against the background if it were just sitting in space. But if it were close to a quasar, the light from the quasar might illuminate it. Or, if it were gaseous, it might be ionized by the light from the quasar, and if so, it could be detected.

The third way quasars may be useful is related to their host galaxy. (We saw earlier that all quasars, even the most distant ones, are now believed to be embedded in host galaxies.) By looking very carefully at the distant quasars we may see some evidence of their host galaxy.

"There are important questions in relation to this host galaxy," said Hu. "Where did it come from? When did it form? Is there any material left over from its formation? If there is material left over, you might hope that the quasar is illuminating it. It is quite possible."

Hu is using the University of Hawaii's 88-inch telescope and the Canada–France–Hawaii telescope in her research, along with the latest CCD arrays. She is also hoping to use Keck when it comes on line.

OTHER SURVEYS

Olivier Le Fèvre of the Canada–France–Hawaii Corporation and F. Hammer of the Paris-Meudon Observatory in France have also been working on a deep space project. They have observed and studied about 30 objects with a redshift greater than 1.

More recently Le Fèvre says he has become interested in the galaxies between us and a redshift of 1. Just as Cowie's group does, he picks a random area of the sky and counts all galaxies in the area, and determines their redshifts. On any given plate there will normally be a distribution of galaxies, some relatively close to us, others more distant, and some as far away as 9 billion light-years. His survey now includes more than 100 redshifts; it differs

Olivier Le Fèvre.

from Cowie's in that he is mainly interested in the galaxies between us and the most distant ones.

"What we are doing right now is very time-consuming in
telescope time," said Le Fèvre. "Just to get the spectrum of one
galaxy at a redshift of 1 can take up to 8 hours. If you want 1000,
a lot of time is needed." He has been getting around this by using
a multiobject spectrograph (MOS). In one exposure you can get the
redshifts of up to 100 galaxies at once with this instrument.

"We know the luminosity function of galaxies around us, and

we want to see how it changes as you go back to redshift 1," said Le Fèvre. I asked him how evolution affects galaxies. "There are several things that might make galaxies brighter in the past—starbursts, for example," he said. "But you also have merging of galaxies which would produce bigger and therefore brighter galaxies as time goes on. Both effects are important. So galaxies may not be brighter in the past." He said his observations hadn't confirmed it one way or the other.

The barriers near the boundary of the universe are being pushed back farther and farther, and considerable progress is being made by observers at Mauna Kea. We may, indeed, find the "end of the universe" in the next few years.

CHAPTER 12

Stars and Stellar Debris

From the summit of Mauna Kea the heavens are filled with stars. The Milky Way stretches in all of its splendor from horizon to horizon. Stars parade overhead in endless patterns and variety, season after season, yet strangely the study of stars gets less publicity than the study of quasars, black holes, and cosmology. Still, it is a central part of astronomy, involving hundreds and even thousands of astronomers. Much is known about stars, but much is still to be learned.

EARLY STELLAR ASTRONOMY

One of the major tools in the study of stars is the spectroscope. With it astronomers have brought about a celestial revolution; the mystery of the stars has been revealed through groups of innocent-looking lines. These lines are like fingerprints; they tell us an incredible amount about the stars: their true brightness, their distance, the temperature of their surface, the composition of their atmosphere, and much more.

In the late 19th century Edward Pickering of Harvard University developed a spectroscope that allowed him to photograph the spectrum of large numbers of stars with one exposure. The most prominent lines in most of the stars (assuming they were not too cold or hot) were those related to hydrogen. Pickering saw that the strength of these lines varied considerably from star to star, and he decided to use them to classify stars. The most intense lines according to his scheme were designated as type A, the second most intense as type B, and so on. For a few years the system seemed to work, but then one of Pickering's helpers, Annie Jump Cannon, noticed peculiarities in the lines of other elements. Something was wrong. A new, revised scheme was needed, but the letters that had been used to classify the various groups had become established in the literature by now, and it was difficult to change them. Pickering and Cannon therefore left them as they were and rearranged them so there was continuity in the lines. The new sequence became O, B, A, F, G, K, M. It was soon shown that this new sequence correlated with surface temperature, with O-type stars the hottest and M-type, the coolest. This scheme is still used today.

As astronomers continued to study stars, spectra became an increasingly valuable tool. It was hard to believe how much information was packed in a few little lines. Not only did a spectrum allow astronomers to determine the absolute brightness, or luminosity, of a star, but it also gave the star's surface temperature with considerable accuracy. Two astronomers decided to take advantage of this information at about the same time (1905): Ejnar Hertzsprung in Denmark and Henry Norris Russell in the United States. Just as someone might decide to plot a graph of height versus weight for a group of people, Hertzsprung and Russell decided to plot luminosity versus surface temperature for the stars around us. They found that most stars lay in a broad diagonal strip across the plot. You would get the same result if you plotted height versus weight for a large group of people, since most people are close to the average weight for their height. Of course, some people are extremely heavy for their height, and others very light. They would not be on the diagonal. In the same way Hertzsprung and Russell found that a few of the stars did not lie on their diagonal (we now refer to this diagonal as the main sequence). Some of the stars were in the upper right-hand region of the plot, and a few were in the lower left-hand region. The stars in the upper right were shown to be red giants; those in the lower left, dwarf stars.

The stars in the main sequence itself varied in color and size. At the lower right there were tiny red dwarfs. Yellow stars like our sun were about halfway up the column, and at the top were hot, massive blue and white stars. This diagram was eventually called the HR diagram in honor of Hertzsprung and Russell.

Arthur Eddington of Cambridge University in England became intrigued with the HR diagram. He was sure it was the key to the internal structure of stars, and perhaps their radiant energy. At this time very little was known about the interior of stars. How hot were they? What were they composed of? The Swiss astronomer Robert Emden had suggested that they might be gaseous throughout, but most astronomers thought they consisted of some sort of hot liquid.

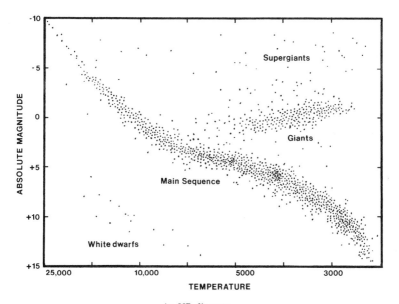

An HR diagram.

A mathematical genius and child prodigy, Eddington won honor after honor in school, capping it off with a scholarship to Manchester University. Not only was he at the top of his class, he was also the youngest (15) when he graduated. But when he went to register at Manchester he was surprised to find they wouldn't admit him: he was too young. After reviewing his case, however, officials quickly relaxed the rules. From Manchester Eddington went to Cambridge for graduate work. In 1906 he became chief assistant at the Greenwich Observatory. In 1913 he returned to Cambridge, and a year later he was appointed director of the Cambridge Observatory. He was barely 30.

At about this time Eddington began working on the structure of stellar interiors. He was soon convinced that Emden was right: stars were gaseous throughout. But what was needed for equilibrium? Gravity would, of course, pull the gas of the star inward, and this inward pull would have to be balanced by an equal outward force. Eddington showed that the gas and radiation pressure (assuming the gas was exceedingly hot in the interior) would exert an outward force and could counter that caused by the inward gravitational pull; this allowed him to calculate the temperature at the center of the star. He got 40 million degrees. We now know that for most stars, it is only about half that.

In 1926 Eddington published his results in his classic book *Internal Constitution of the Stars*. The basic ideas and equations of stellar structure were given, but something critical was missing. What was causing the tremendously high temperatures at the core of the star? They had to be exceedingly high to balance the inward gravitational pull, but Eddington was not sure how they were created. There was, however, a clue. In 1905 Einstein published an equation showing an equivalence between mass and energy, and if mass could be converted to energy, the amount of energy created would be enormous. This meant that if several hydrogen nuclei fused to produce a helium nucleus (of less mass), a tremendous amount of energy would be released. Eddington eventually came to the conclusion that this was, indeed, how the energy of the stars was generated. The idea was particularly satisfying in that stars were made up mostly of hydrogen, and there would be an ample supply of fuel. The exact conditions under which the fusion took place, however, were not known.

If two hydrogen nuclei (protons) are to fuse they have to get exceedingly close to one another. This is difficult because both have the same electrical charge, and like charges repel. They can only get close to one another if they collide at tremendous speeds—speeds that are only possible at very high temperatures. In Eddington's time 40 million degrees was not considered to be nearly high enough.

A breakthrough came in the late 1920s when theorists in Germany showed that the "billiard ball" approach of Newton wasn't valid in the realm of atoms. Atoms didn't act like billiard balls when they collided. To determine what happened you had to use quantum mechanics.

Europe was soon a hub of scientific activity centered on quantum mechanics. The best scientific minds in the world were attracted to the major universities at Göttingen, Berlin, and Munich, where the discoveries were being made. One of them was George Gamow, who came from Russia to learn the new theory. To his dismay, however, he found that so many people were working on the foundations of the theory, it was almost impossible to make a significant contribution. He turned, therefore, to one of the well-known problems of physics: radioactive decay. Could the new mechanics be used to explain the spontaneous decay of heavy elements? It was well-known that heavy elements such as radium decayed to lighter ones with the release of alpha particles (helium nuclei). Gamow solved the problem using quantum mechanics, and was surprised with what he got. According to traditional (classical) physics, the alpha particle couldn't get out. The heavy radium nucleus was surrounded by a huge impenetrable barrier— almost as if a mountain were surrounding it. Gamow showed that according to quantum mechanics, there was a small probability that the alpha particle could penetrate the mountain. It need not jump over it; it could "tunnel" through. It was a rare phenomenon but occurred occasionally, and that was all that was needed to

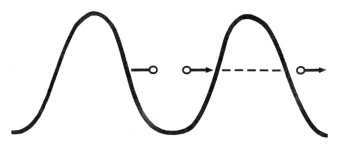

Schematic of quantum tunneling through a barrier.

explain the phenomenon. The new effect was called the tunnel effect.

But if heavy particles could tunnel out, it was equally possible a particle could tunnel in through the barrier. Could two hydrogen atoms, for example, fuse via this mechanism? The physicist Robert Atkinson and Fritz Houtermans decided to look into this in 1929. And sure enough, they found it was possible; furthermore, it was possible at a temperature of 15 million degrees, which was believed to be the temperature at the centers of stars.

Atkinson and Houtermans proved that nuclear fusion could fuel the stars; hydrogen could be converted to helium via quantum tunneling. Houtermans was so pleased with the discovery he took some time off while writing it up to go for a walk with a young lady. As it got dark she looked up at the stars and said, "How beautiful they twinkle." Houtermans, still on a high from the discovery, proudly boasted, "Yes, and I'm one of the only persons on Earth that knows why they shine." He was extremely disappointed, however, when she didn't seem the least bit impressed.

A significant breakthrough had been achieved, but the details of how hydrogen fused to produce helium were still unknown. It was not a simple process; two hydrogen nuclei didn't just collide to give helium. It was a complex process, and ten years would pass before anyone discovered it.

The discovery was made by Hans Bethe. Born in Strasbourg, at that time part of the German Empire, but now part of France, Bethe grew up in Kiel and Frankfurt, attended the universities of Frankfurt and Stuttgart, then later went to the University of Munich. He was working on his Ph.D. at the time quantum mechanics was being formulated, and he quickly became an expert in the new theory.

In 1930 Bethe was awarded a scholarship to study at the Cavendish Lab in England, and with Enrico Fermi in Rome. He returned to Germany in 1932, accepting a position at the University of Tubingen. Within a short time, however, Hitler came to power, and Bethe, who was part Jewish, lost his job. He emigrated to England where he taught for a short time at Manchester and Bristol Universities; then, in 1935 he accepted a position at Cornell University in the United States.

In 1938, while attending a physics conference at the Carnegie Institution in Washington, D.C., his attention was drawn to the problem of the energy of stars. Why do they shine? Hydrogen had to be converted to helium—but what were the exact steps? He began working on the problem on the train ride back to Cornell, and within a few weeks he had the answer. A series of nuclear reactions, with carbon acting as a nuclear catalyst, would produce the desired results. This series is now called the carbon cycle. The same cycle was discovered independently by Carl von Weizsäcker in Germany at about the same time.

A short time later Bethe discovered a simpler cycle, now called the proton–proton cycle, that also converts hydrogen to helium. It is the cycle that is presently operating in our sun. The carbon cycle operates in slightly hotter stars. Bethe was awarded the Nobel prize for his discovery in 1967.

THE LIFE CYCLE OF A STAR

Eddington had given the equations that would allow astronomers to look inside a star. But to create a model of a star using these equations was a tedious task, requiring many hours. Typically, an astronomer would begin by dividing the star into shells, or layers. Then, using the known temperature, luminosity, pressure, and mass per unit volume at the surface, along with Eddington's equations he could determine the same properties in the layer just below it. Once these quantities were obtained he could use them to calculate the corresponding ones in the layer below it, and so on all of the way through to the center.

It was a job few cared to tackle. But when computers came on the scene it became relatively easy. Furthermore, with computers it was possible to determine how the model changed as the star aged. This presented astronomers with a tremendous opportunity; they could use computers to learn about the evolution of a star.

The first step was to find out what happened to the star when it had used up most of its fuel, namely hydrogen. The center of the star at this stage would be made up of helium, the "ash" left from the "burning" of hydrogen. Being heavier than hydrogen it would sink to the center.

Fred Hoyle of Cambridge University and Martin Schwarzschild of Princeton University set out in the early 1950s to find out what happened when the center of the star became contaminated with helium. Both men had made important contributions to astronomy. Using computers they were able to show that stars like the sun gradually bloated and became red giants.

Within a few years other astronomers began using computers to learn about stellar evolution. How was a star born? What happened to it shortly after it was born? What happened to it in old age? Computer technology developed rapidly during the 1950s and 1960s, but larger and faster computers were not all that was needed. Better computational techniques were also needed, and in 1961 Louis Henyey of the University of California developed a new technique that allowed astronomers to build models much more efficiently. This technique along with faster computers enabled astronomers to follow the life cycle of a star in considerable detail. The computers told them how a star was born, and what happened to it as it aged.

The computer models were compared with observations and the agreement was good. We can't, of course, see stars age. Even over many human generations they change very little. But we do see stars at different stages through their life, and we can compare our models with these stars.

Summarizing the results, we find that a star is formed from a gas cloud. This gas cloud is composed mostly of hydrogen and helium, but other elements are present in tiny amounts (less than 1 percent). Self gravity pulls the cloud inward, increasing the pressure and therefore the temperature at the center until finally the cloud becomes a hazy red gaseous sphere. The temperature at the center continues to climb as gravity continues to compress it. It passes 1 million degrees, then 10 million; finally it reaches 15 million degrees and nuclear reactions are triggered. We usually refer to this as "burning"; hydrogen is burnt, leaving helium as ash. The helium, being heavier than the hydrogen, piles up at the center, and soon there is a helium core. The buildup doesn't stop here, however; pressures continue to climb, and the temperature at the center of the helium core now skyrockets; when it reaches 100 million degrees the helium is ignited. In an average-sized star like the sun, however, helium is explosive, and the entire core is blown apart. The hydrogen burning, which is now occurring in a shell around the helium core is ripped apart, burning is extinguished, and the nuclear furnace goes out. This is called the helium flash.

The star at this stage is so extended—its outer shells have been bloated by the high temperature—that we see little evidence of the explosion at the surface. Within a few years, however, the star begins to dim; it continues to dim rapidly for several years. Gradually, though, the helium falls back to the center and burning begins again—peacefully this time. The shell of burning hydrogen around the helium also re-forms, and the star begins to brighten; soon it is back to its original magnitude.

The star is now producing energy by burning helium at its core, and hydrogen in a shell around the helium. But the burning of helium also produces "ash." The ash in this case is a mixture of

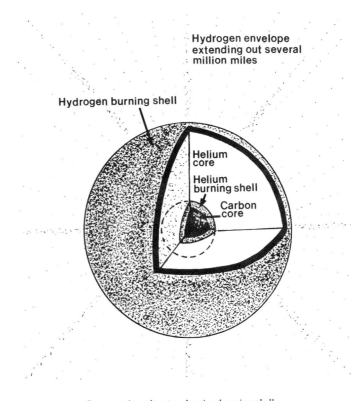

Cross section of a star showing burning shells.

carbon and oxygen, and since it is heavier than helium, it settles at the center. Helium burning continues in a shell around this new core, and the temperature at the center continues to climb. Then the carbon is ignited, and again it leaves ash. This continues until the star is burning several elements in shells, one enclosing the other. If the star is massive enough, it can produce elements all of the way up to iron. Iron, however, is the end of the line; fusion cannot produce heavier elements. If the temperature continues to climb the entire star is blown apart as a supernova.

Supernovae can also occur in less massive stars, but they do not occur in stars like our sun. Our sun will burn helium, but it will not generate temperatures high enough to burn anything further. After burning helium it will slowly collapse over millions of years, ending as a small dense object called a white dwarf. In the HR diagram white dwarfs are in the lower left-hand corner.

The life cycle of a star, as just described, can be followed using the HR diagram (see figure). In this diagram the star is formed in the upper right. It moves slowly toward the main sequence; at the main sequence nuclear reactions are triggered. A star spends most of its life in the main sequence, but eventually its fuel (hydrogen) is depleted and it moves to the upper right where it becomes a red giant. The helium flash occurs in this region. The star stays here as it continues to generate heavier and heavier elements. Finally it

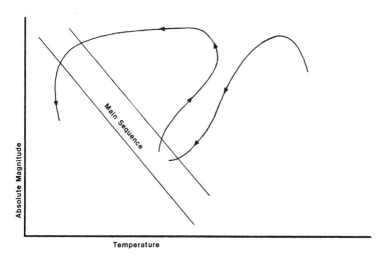

Lines and arrows indicate path of star during lifetime. It moves to the main sequence from the right, then later moves to the upper right and finally to the left and down.

moves to the left, passing through the main sequence. Its final collapse takes it to the lower left; if it is massive enough it will supernova. Otherwise it will collapse slowly and become a white dwarf.

HERBIG STARS

Many astronomers are studying stellar structure and evolution using the telescopes on Mauna Kea. One of them is George Herbig of the Institute for Astronomy, whose main interest is young stars. In recent years he has also been studying the material from which stars are built—the interstellar medium.

George Herbig.

Herbig was born in a small town in West Virginia. He went to UCLA, then on to the University of California at Berkeley for his Ph.D. where he did a thesis on young stars called T Tauri stars. They are stars about the mass of the sun that have high outflowing winds and are found only in dark nebulae. "There wasn't anyone at the University of California at the time I was there who was interested in young stars, so I didn't really have a thesis director," said Herbig. "I was sort of self-propelled." He graduated in the late 1940s, took a one-year postdoctoral in the East, then accepted a job at Lick Observatory on Mount Hamilton, which later became part of the University of California at Santa Cruz. He stayed at Lick until 1987 when he came to the Institute for Astronomy in Hawaii.

Herbig has had the honor of having two different kinds of stellar objects named after him. He discovered the first kind, now called Herbig–Haro objects, while working on his thesis. Some of the young stars he was studying had strong winds blowing out of them. Herbig noticed that clumps of gas were sometimes picked up by this wind and carried away at high velocities. If these clumps struck the ambient gas surrounding the star (the gas from which the star formed) a shock front formed on one surface. This shocked gas had a unique spectrum that could easily be distinguished from other types of hot gas.

Herbig is also known for having recognized what are called Herbig stars. In 1960, about 10 years after he discovered the Herbig-Haro objects, his attention was drawn to the fact that although newly forming stars about the mass of the sun had been identified (they were the T Tauri stars), their counterparts at larger masses—5 to 10 times that of the sun—had not been found. "I looked around and came up with a list of about two dozen stars that behaved according to expectation," said Herbig. "They are large hot stars of spectral types A and B that exhibit bright spectral lines."

A considerable amount of work has been done on Herbig stars in the last few years, but many details are still not understood. One of their most interesting properties is their light variability. Although some of the changes are small—only a fraction of a magnitude—occasionally they vary by several magnitudes. What causes this change? Because they are particularly bright in the infrared, astronomers have come to the conclusion that, in addition to gas, they are surrounded by shells of dust and gas. The clumpiness of this dust, as it orbits the star, causes the light variations. There is a problem with this, however: this material orbits the star in an equatorial disk, and we would have to see this disk edge-on. The likelihood of this is not great, and because of it, other models have been suggested, the most popular being one in which some of the changes are caused by emission from the star's atmosphere.

I asked Herbig about the problem. "I think there's no doubt there's a heavy concentration of dust somewhere nearby," he said. "And I believe the variations have to be associated with the dust moving across our line of sight. If it was something the star itself did to get fainter by cooling off, then it would also get redder. But in many cases it doesn't." A recent observation, in fact, showed that the temperature of a Herbig star did not change while its light decreased by 90 percent.

Herbig is working on two projects at the present time. One is related to Herbig stars. "There's a loophole in this business," he said. "Everyone is content with our idea of how a 5-solar-mass star looks when it is young. But in a region like the Orion nebula where there is a whole range of young stars, you find some massive stars that don't act like the Herbig stars are suppose to act. So there could be another channel through which stars of this kind get to the main sequence. I am looking into this."

Herbig is also working on the interstellar medium. He became interested in it through his fascination with young stars. This is the material from which stars are built, and Herbig felt that he had to understand more about it, if he was to understand how young stars came about. He is particularly interested in the spectrum of this material. We don't, of course, see the spectrum of the interstellar medium directly, but it produces absorption lines in the spectra of stars seen through it. Many of these lines are recog-

The Orion nebula. (Courtesy National Optical Astronomy Observatories)

nizable, but a number of broad, diffuse bands are not; they are referred to as the diffuse interstellar absorption bands. They were found in the 1920s, but are still not thoroughly understood. None of these bands can be identified with materials produced in the laboratory.

"The chemistry of simple interstellar molecules is fairly well understood," said Herbig. "We understand where CO, CN, and so on come from. They're formed in reactions involving molecular hydrogen. You would therefore expect that the unknowns here would just be more complicated molecules associated perhaps with H_2. It turns out, however, that they're not. They're correlated with the abundance of neutral hydrogen atoms, so they must come about in a completely different way." The major question is therefore: what is this material and where does it come from? Herbig is still uncertain. He mentioned, however, that several groups are now trying to produce these molecules in the laboratory, so far without much success.

LOOKING FOR LITHIUM

One of Herbig's former students, Ann Boesgaard, of the Institute for Astronomy, is also working on stars. She is particularly interested in their lithium and beryllium. The abundance of these elements can tell us quite a bit about their internal structure.

Born and raised in Rochester, New York, Boesgaard developed an early interest in science and mathematics. She mastered the multiplication tables in kindergarten, and by the second grade she could identify many of the planets and knew some of the constellations. Science was her second favorite subject, she said, ranking right behind gym.

Upon graduation from high school she went to Mount Holyoke College where she majored in astronomy, physics, and mathematics. Astronomy soon became her favorite. Occasionally she needed to observe late at night, but there was a problem: an 11:00 P.M. curfew. Since it was a lot of trouble to get special permis-

Ann Boesgaard when she graduated (Ph.D.).

sion, she would occasionally sneak past the campus police and make her observations, then sneak back to her room.

By now she was determined to become an astronomer, and since most of the astronomical telescopes were in California she decided to go to the University of California for graduate work. When she told her advisor at Mount Holyoke about her ambition, however, she was told that a woman would never be allowed to use the telescopes. Undeterred, she went to the University of California, and soon had her first look through a large telescope—the 120-inch reflector at Lick Observatory.

Her Ph.D. thesis was on lithium and beryllium in stars. Lithium and beryllium are important because, unlike most other elements in stars, they are easily destroyed by nuclear reactions within the star. For lithium it takes only a temperature of 2 million degrees—relatively low compared with the 15 million at the core. For beryllium it is approximately 3 million degrees.

Boesgaard completed her Ph.D. in 1966 and went to Caltech on a postdoctoral fellowship. She took spectra with the 100-inch Hooker reflector at Mount Wilson, the telescope that Edwin Hubble had used to show that the universe was expanding. "I loved observing with that venerable telescope," she wrote. "It was so quiet and peaceful with strains of 'Music to Dawn' coming over the radio. I was delighted to have the opportunity to collect so much data and to learn so much about stellar spectroscopy."

Boesgaard married soon after receiving her Ph.D., but her husband, Hans Boesgaard, was now in Hawaii helping build the 88-inch University of Hawaii telescope. She made several trips to Hawaii to visit him, and on one of them she was offered a position at the University of Hawaii. She was, in fact, one of the first people brought in by John Jefferies to help build up the astronomy program. The University of Hawaii had just been given the go-ahead on the 88-inch telescope. "My husband and I went to the summit of Mauna Kea before there was anything there," she said. "There were no telescopes. They were just flattening off the top of the cinder cone for the foundation of the 88-inch telescope dome."

The early years at the University of Hawaii were exciting ones for her. "It was really exhilarating most of the time because we were doing this pioneering . . . making a telescope on a mountain where people said you couldn't breathe. We found out early on you could."

The 88-inch telescope was dedicated in 1970. Boesgaard had the honor of taking the first spectrogram with it. It was the spectrum of our nearest star, Alpha Centauri. "Alpha Centauri is a southern star, so it is very near the horizon in Hawaii," said Boesgaard. "So there was something nicely poetic about showing off the field of view of our telescope by looking that far south."

The second winter after the telescope was dedicated there was a heavy snowstorm that left a considerable amount of snow and ice on the summit. She and Alan Stockton were scheduled to observe the night after the storm, but the top of the dome was covered with ice and they couldn't get the shutter open. They therefore climbed to the top of the dome along a ladder on the outside with ice picks and knocked the ice off. "We had safety belts on but it was pretty scary up there," she said. "They wouldn't allow us to do that now." She laughed. "In fact, John Jefferies probably wouldn't have allowed it then—if he had known."

In 1990 Boesgaard was awarded the Muhlmann Prize by the Astronomical Society of the Pacific for work she started in 1986 with graduate student Michael Tripicco. They looked at the lithium abundance in F stars in the Hyades Cluster. F stars have a slightly higher surface temperature than our sun, and are about 25 percent more massive. Their temperature ranges from 5900 to 7000 K.

"We expected to see that all these stars had about the same abundance of lithium, but were surprised to find there was a narrow range of temperatures, extending about 400 degrees, where the lithium content plummeted. It had to have been circulated to lower layers where it was destroyed," said Boesgaard. Many stars, depending on their mass, have a convection layer just inside the outer surface. And just as convection currents circulate heat in the home, these convective currents circulate heat from a layer just below the surface, to the surface itself, then back down again. Any lithium on the surface would be transported to the hotter layers below, and if these layers were above 2 million degrees, the lithium would be destroyed. Any lithium in the outer atmosphere of the star would therefore disappear. Lithium can therefore be used to trace the depth to which convection penetrates.

Astronomers have discovered that there is a critical temperature (slightly higher than the sun's surface temperature) above which are no convection currents in stars. Lithium is present in its normal abundance in these stars. Boesgaard and Tripicco discovered, however, that in a certain range where there should have been lithium, there was none. They referred to it as the "lithium gap."

Interestingly, this gap occurs just below the critical temperature for convection in stars, and above this temperature, stars rotate very rapidly. Rotation is connected with convection because convection regions have very strong magnetic fields associated with them, and these magnetic fields act to "break" the rotation of the star, slowing it down. Because of this, cooler stars rotate at a lower rate.

Boesgaard and Tripicco found that rotation decreased significantly across the lithium gap. Boesgaard believes the gap can be explained if convection currents are not the only mechanism regulating lithium abundance. Theorists had shown that rotation can produce an additional circulation called meridional circulation. Boesgaard believes that this, together with convection currents, carries lithium to even greater depths, and consequently destroys lithium more effectively than convection currents would acting alone. In short, lithium declines from 7000 to 6700 because the convection zone gets larger and meridional circulation helps carry it to greater depths. The lithium abundance rebounds from 6700 to 6200 because stars rotate much more slowly as the temperature decreases.

Much has been learned about stars by the astronomers of Mauna Kea, and the study of stellar properties will continue to occupy a central role on the mountain.

CHAPTER 13

Searching for Other
Planetary Systems

How often have you wondered if there exists, somewhere among the tiny points of light that dot the night sky, a planet similar to ours that sustains life? It would, indeed, be the most exciting discovery ever made. Think of the implications. What would we learn from it? The discovery of even one planet with life out there would imply that the universe is full of life.

But even if we did detect a planet with a civilization, communicating with it would be extremely difficult. We are restricted by the speed of light. Even in the case of a planet orbiting our nearest star, we could not send a message and get a reply in less than nine years. And we're quite certain there's no life on it. For a star 30 or 40 light-years away, which, astronomically speaking, is extremely close, it would take 60 to 80 years to get a reply. Communication with an extraterrestrial civilization is therefore going to be difficult. So for now, let's concentrate on the planets themselves—extrasolar planets.

EARLY SEARCHES

Much of the early interest in extrasolar planets centered on the dim red star known as Barnard's star. Only six light-years away, it is so dim we cannot see it with the naked eye. It has a magnitude of approximately 10 (with the naked eye we can only see to about magnitude 6). Yet, strangely, it is our second nearest system; only the triple system, Alpha Centauri, is closer.

Barnard's star was discovered by Edward Barnard of Lick Observatory in 1916. After photographing a section of the sky in the constellation Ophiuchus, almost as an afterthought he compared it to a photograph he had taken of the same region about 20 years earlier. To his surprise he found that one of the stars had moved much more than he thought possible. He measured its proper motion (angular velocity across the sky) and was amazed to find that it was the largest ever recorded.

The star came to Peter van de Kamp's attention in 1938. A year earlier van de Kamp had become director of the Sproul Observatory, at Swarthmore College, near Philadelphia. One of his first acts as director was to initiate a search of nearby stars for planetary companions. He knew there was little chance of detecting planets directly, but a method of detecting them indirectly had been used by Friedrich Bessel of the Konigsberg Observatory in Prussia in the mid-1800s to show that Sirius had a tiny companion.

Van de Kamp knew that if a star had a planet, the star would appear to "wobble" as it moved across the sky. To see why, consider the Earth. We know that the Earth orbits the sun. This doesn't mean that it revolves around the sun's center. If you look closely you see, in fact, that it doesn't. The two objects revolve around their center of mass. This is the point where they would balance if the Earth was placed on one end of a teeter-totter and the sun on the other end. The sun is thousands of times heavier than the Earth, so the balance point would be well inside the sun. If Jupiter were placed on the teeter-totter with the sun, however, the balance point is just outside of the sun.

What is important here is that when a star and its planet move through space they are revolving about their center of mass. At the same time they are tracing out a much larger orbit around our galaxy. If we looked at the system through a telescope we wouldn't see the planet, but we would see the star "wobbling" as

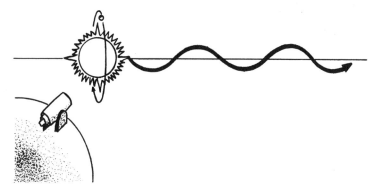

Schematic showing wobbly motion of a star through the sky caused by a dark object orbiting it.

it moved through space. This wobble would be the result of the planet tugging on it, and it would be a clear signature of its presence. To see the true extent of the wobble, you would have to observe it over a relatively long period of time—usually several times its orbital period. For Jupiter and the sun that would be 20 to 30 years.

Van de Kamp began photographing Barnard's star in 1938, and over the next few years he photographed it 30 times a year. Using the photographs he plotted its path across the sky. After a few years there was some deviation from a straight line, but not enough for him to be sure. He continued plotting its path for more than two decades; finally in 1962 he was convinced there was enough evidence to announce his findings. There was, according to his measurements, a planet 1.6 times as massive as Jupiter orbiting Barnard's star, and it had a period of 24 years.

It was a startling announcement. No one had ever detected a planet before. But van de Kamp wasn't finished. He continued taking data, and seven years later he announced that a system of two planets fitted the data better than a single planet. According to his calculations a planet with a mass approximately 70 percent that of Jupiter orbited with a period of 12 years. Further out, one with about half Jupiter's mass orbited with a period of 24 years.

If van de Kamp's discovery was verified by other astronomers, it would be a momentous discovery. The implications were enormous. Barnard's star was one of the nearest stars in the sky, and if it had two planets, most other stars also likely had planets. Several astronomers began checking to see if they could, indeed, verify his results. One of them was George Gatewood of Allegheny Observatory. He gathered plates from several observatories that showed Barnard's star. Using them he plotted its position over several years, and although he found a small deviation, it was insignificant compared with van de Kamp's. He announced that van de Kamp's results appeared to be wrong.

Robert Harrington, a former student of van de Kamp's, who was at the United States Naval Observatory, also began looking into the problem. Using a 61-inch refractor at the Flagstaff Observatory in Arizona, he photographed Barnard's star for several years, and was also unable to verify van de Kamp's results. He found a slight deviation from a smooth curve, but nothing to indicate there were two large Jupiter-sized planets orbiting Barnard's star.

Gatewood soon started his own observing program, and after a few years he found that his data also contradicted van de Kamp's. Interestingly, both Gatewood and Harrington found small perturbations in the orbit; in particular, both recorded a small discontinuity in the curve in 1977. So planets can't be ruled out, but it now seems unlikely they are as large as Jupiter, if they exist.

Van de Kamp's results became even more controversial when John Hershey of Sproul Observatory showed that 12 other stars measured at Sproul had a wobble very similar to Barnard's star, and significant deviations had occurred in all stars in 1949 when a new cell for the telescope was installed, and again in 1957 when further adjustments were made.

BASIC TECHNIQUES

The technique that van de Kamp, Harrington, and Gatewood used is indirect. They did not see the planet, but noticed its effect on the star. Until telescopes become much more powerful, and instrumentation much more sensitive, this is likely the only technique that will be used. It is referred to as astrometry by astronomers.

For astrometry to work well the planet has to be relatively massive. The larger the planet's mass compared with the star's mass, in fact, the larger the wobble. The Earth causes almost no wobble in the sun's motion. Jupiter, on the other hand, causes a wobble roughly equal to the sun's diameter, and it could be measured from nearby stars with a large telescope.

Not only must the planet be relatively massive but you must have good reference stars. If you are to plot the star's motion through the sky accurately, you need to compare its position with the stars around it, and this can be a problem, as all stars move to some extent. It is important, therefore, to have as many reference stars as possible.

In the technique that van de Kamp used, only one component of the star's three-dimensional motion through space was observed. Referred to as the tangential component, it is perpendicular to our line of sight. There is, however, another component called the radial component; it is along the line of sight and can be either toward us or away from us. This component, it turns out, is much easier to measure than the tangential component. It can be obtained using a spectrograph. Because of its wobble, the star will move back and forth along our line of sight. The spectral lines from the star will therefore shift toward the red end of the spectrum (when the star is moving away from us), and toward the blue end (when it is moving toward us). This shift can easily be measured.

The spectroscopic method has another advantage over the astrometric method. It's easy to see that the farther a system is from us, the smaller its wobble will appear. In fact, a system must

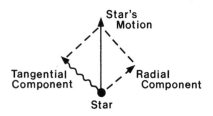

Radial and tangential components of a star's motion.

be relatively nearby for its wobble to be detectable. This is not necessarily the case for a spectroscopic system. As long as there is enough light to obtain the spectrum, the shifts can be measured. Much more distant systems can therefore be examined.

Just as we need good reference stars in the astrometric system, we need good reference spectral lines in the spectral method. The amount of shift is usually small, and accurate measurements are therefore needed. Excellent results have been obtained by placing a gas absorption cell in front of the entrance slit of the spectrograph. The spectral lines from the gas are superimposed on the spectra from the star, and they remain stationary while the star's lines move. Two different gases have been used: hydrogen fluoride and iodine.

The other method of detecting planets is, of course, observing them directly—in other words, detecting the photons they reflect. This is not easy, as stars are usually billions of times brighter than

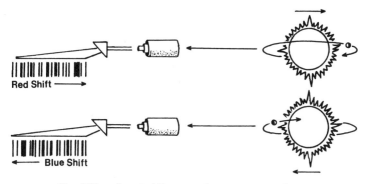

The shifting of spectral lines as a planet moves around a star.

planets. A planet's feeble light is therefore "swamped" by the light of the star.

Once planetary candidates have been detected indirectly, astronomers will no doubt attempt to observe them directly. And there is a method for eliminating the light from the star that they orbit. The technique was first used in the 1930s by Bernard Lyot of France to create an artificial eclipse of the sun. He constructed a device, called a coronograph, using a small circular disk to block the light of the sun. In the same way we can blot out the light from the star using a tiny disk on the optic axis of the telescope. Interestingly, the disk in this case is no larger than one of the dots on this page. There is a problem with this method, however. Light still gets through around the edges because of optical imperfections, and in most cases it obscures the planet. We will see later, however, that this technique is useful in searching for disks of debris that may have formed around a star.

THE CFH–DAO PROGRAM

One of the most successful spectroscopic searches for extrasolar planets is one that was initiated in the late 1970s by Bruce

Campbell and Gordon Walker of the University of British Columbia; Campbell was a student of Walker's at the time. Most of the early work was done by Campbell while he was a resident astronomer at the Canada–France–Hawaii telescope. David Bohlender, one of the current resident astronomers at the CFH telescope, is now actively involved in this program. I caught up with him at his office at CFH headquarters in Waimea.

Born in Chesley, Ontario, Bohlender received his bachelor's degree in astronomy from the University of Western Ontario. He developed an early interest in astronomy, but by the time he finished his degree, he was beginning to have second thoughts about it. "I was getting a little disenchanted," he said. "The accu-

David Bohlender.

racy of measurements and so on, seemed so poor, and there were so many untested ideas." After giving it some thought he decided to switch to something else. He moved to the University of Calgary and registered for a master's degree in atmospheric physics. By the time he had finished this degree, however, he found he was beginning to miss astronomy. Furthermore, it seemed that the accuracy of things in atmospheric physics was even worse than in astronomy. He decided to go back to the University of Western Ontario for a Ph.D. in astronomy. His thesis, which was on stars with strong magnetic fields, was done under John Landstreet.

While at the University of Western Ontario he used their 40-inch reflector extensively, but it wasn't large enough to obtain the data he needed for his thesis, so he applied for time on the CFH telescope. He was allotted six nights in 1986. The first night was so cloudy and miserable it scared Bohlender. "I was sure the whole run was going to be washed out," he said. "I thought I'd have to select a new thesis topic. I was really worried. But then we had five great nights in a row and I got beautiful data."

The goal of his thesis was to determine the surface abundance of several elements and determine the strengths of the magnetic fields in a number of magnetic stars. He got so much data during the run that six years and several publications later, he was still working on it. He has used the CFH telescope several times since, but when I talked to him it had been several months since he had a run on it. "I always tell people that this time of year [it was winter when I interviewed him], when Orion is high in the sky, and I don't have any telescope time, I'm kind of glum, because most of the interesting magnetic stars are in Orion."

After receiving his Ph.D., Bohlender was awarded a fellowship that allowed him to go anywhere he wanted to continue his studies. Gordon Walker of the University of British Columbia was doing some interesting spectroscopic work, so Bohlender decided to go to UBC. He spent two years there working with Walker, and traveling to Victoria to observe with the telescopes at the Dominion Astrophysical Observatory (DAO).

About the time Bohlender came to UBC, Bruce Campbell

dropped out of the planetary program. Walker invited Bohlender to join, and although he managed to continue his work with magnetic stars, Bohlender soon became actively involved in the planetary group. Also in this group were Stevenson Yang and Alan Irwin of the University of Victoria. Walker's role decreased significantly after he was asked to organize the Canadian end of the Gemini project in the late 1980s. Gemini is a large telescope that is to be built on Mauna Kea by 1998.

Most of the measurements in the Canadian program are now being made using the CFH telescope. "The giants are bright enough that we can use the DAO telescope in Victoria," said Bohlender. "But for the dwarfs the CFH telescope is essential." He went on to say that there are 23 dwarf stars in the program and about a dozen giant stars. The dwarfs are all less than 30 light-years away, the giants generally much farther. The object of the program is to measure very small velocity changes over long periods of time. Bohlender said that they can now detect velocities as low as 20 meters per second.

Early in the program interest began to center around one of the giants, called Gamma Cephei. In 1987, Bruce Campbell and his colleagues announced that their results indicated that a planet approximately 1.7 times as massive as Jupiter appeared to be orbiting Gamma Cephei; its period was 2.7 years. At the time they had about six years of data. They also announced that they had weaker data for low-mass companions around five other stars.

By 1992, however, there was some doubt about their results. The velocities they were measuring were of the order of a few tens of meters per second, and this was typically what a planet orbiting the star would be expected to produce. But material on the surface of the star can move upward, or downward, at speeds much in excess of that, and it is difficult to distinguish it from the overall velocity of the star. "It's not clear now whether the velocity variations are caused by binary motions [a planet orbiting the star], or by an activity cycle of the star," said Bohlender. "You would expect a star similar to the sun to have an activity cycle causing changes in its convection zone or surface features. This could cause an apparent change in the velocity of the star. One thing that points to this being an activity cycle in some of the stars is that the largest changes are occurring in the giant stars, and they're the ones that tend to be more active."

I asked him if this was a disappointment. (After all, the discovery of a planet is much more exciting than the discovery of an activity cycle.) He chuckled. "Well, we don't know for sure that this is the case. Things are still up in the air."

To check on the possibility that the velocity changes might be related to surface features or convective cells near the surface of the star, the team has been measuring temperature changes that occur during the cycle. If the observed velocity is caused, for example, by material accelerating to the surface as a result of an activity cycle, you would expect to see temperature changes. The results, however, are still inconclusive.

Bohlender pointed out that there was also another problem with the best candidate, Gamma Cephei. According to their results the Jupiter-sized companion was very close to the surface of the star—about twice the distance between the Earth and the sun. A gaseous planet this close would soon evaporate because of the heat, he said.

I asked him what he thought the chances were for eventually finding an extrasolar planet. "I think it's quite good," he replied. "If we can get the noise down by a factor of two or so, I think we have a good chance. It's not a short-term project. I'm sure there are planets out there—it's just a matter of detecting them."

So far in the project they've been using a reticon detector, but they plan on changing to a large CCD late in 1993. Bohlender said that it will make a big difference.

DUST AND DEBRIS

Although we can't observe planets directly at the present time, we can observe material—disks of debris—around several stars. The first inkling that there might be disks of this type came with the launch of the IRAS satellite in 1983. In its relatively short lifetime (less than a year), IRAS was able to map the entire sky in the infrared. Thousands of infrared sources were discovered. Many were, of course, dim red stars, but in a few cases the radiation appeared to be coming from dust close to the star—dust that was being heated by the star. This was significant in that our sun, no doubt, once had a ring of dust around it—the precursor of our solar system.

Two astronomers, Brad Smith of the University of Arizona and Rich Terrile of the Jet Propulsion Laboratory (JPL) in California, decided to take a closer look at these stars.

Smith, who is now actively involved in the space program, started out in chemical engineering, receiving his degree from Northeastern University. After graduation he went into the army where he began working with a team that was making a detailed map of the moon. He also spent a considerable amount of time searching for small undetected satellites of Earth. The army believed that if any existed they might be of military value. He was now sure he would stay in astronomy so he began taking graduate courses at New Mexico State University. A few years later he was on the faculty, and a little later he was involved with the space program, working on Mariner, Viking, and Voyager.

Terrile was born in New York City. He graduated from SUNY at Stony Brook on Long Island, and went to Caltech for a Ph.D. His thesis was on infrared observations of Jupiter and Saturn. When he graduated in 1978 he went to JPL where he worked under Smith, who was the leader of the Voyager spacecraft team.

Smith and Terrile decided to use a large optical telescope to take a close look at the IRAS candidates that appeared to have dust rings around them. They went to the Las Campanas Observatory in Chile, equipped the 100-inch telescope with a coronograph—a tiny mask that cut off the light from the star—and looked at each of the candidates. One called Beta Pictoris proved to be of particular interest; it was only 50 light-years away. It appeared to be surrounded by a disk of gas and dust that extended out 30 times farther than the radius of our solar system. A close analysis of the debris showed that it consisted of particles ranging from dust-sized up to stones roughly the size of your fist.

The discovery of this disk is important, since we know that planetary systems such as our solar system form from disks such as this. It is therefore evidence that other systems are forming. Of particular importance, however, Smith and Terrile discovered that as you move toward the star, the disk thins out, and in the region close to the star, a region about the size of the solar system, there is no dust. It is possible that the dust and debris that were once here have condensed to form larger objects—possibly planets. The material we are seeing may be material that was left over when a family of planets was formed.

Smith and Terrile will have access to the Keck telescope when it is completed and they plan to use it to take a closer look at the ring of debris around Beta Pictoris. With Keck's large mirror and the excellent seeing on Mauna Kea they should get a much better view of it.

TOPS

The Keck telescopes will also play an important role in a program called Toward Other Planetary Systems (TOPS). TOPS was formed in 1988 at a workshop on extrasolar planets sponsored by NASA. At the workshop a three-phase program was outlined and presented to NASA officials for consideration. In the first phase, called TOPS-0, a ground-based search will be made for candidates. This will be followed by TOPS-1, a space-based look at the candidates, and search for further candidates. And finally TOPS-2 will be a more extensive search to begin several years in the future; more elaborate spacecrafts will be used along with telescopes on the moon.

The first phase got a boost in 1991 shortly after Howard Keck of the Keck Foundation announced that they would partially fund a second 10-meter telescope, identical to the first. The Keck Foundation would provide $75 million, which would be about 80 percent of what was needed; CARA would look after the remaining 20 percent. The Keck telescope seemed ideal for the first stage of the TOPS program, and for other solar system studies, so NASA decided to pick up the tab for the remaining 20 percent, in return for an equivalent amount of observing time. Some of this observing time will be allotted to TOPS.

The TOPS program is directed at stars similar to the sun, as they are known to be good candidates as life-supporting systems. The nearest 100 stars will be checked. Spectroscopically, the sun is a G-type star. Both G-type and F-type stars (slightly hotter than G types) will be considered as candidates. M and K dwarfs will also be considered because they are ideal candidates for the indirect techniques discussed earlier. Both are less massive than the sun, and if they have a planet, the ratio between the planet's mass and the star's may be relatively large. As we saw earlier they are the systems that produce the largest wobble.

Of particular importance in both indirect and direct methods is resolution. If a star is about 30 light-years away and has a planet about one astronomical unit from it (an astronomical unit is the distance from the Earth to the sun), the resolution needed to distinguish the planet, assuming the glare from the star is eliminated, is roughly 0.1 arc second. This is presently beyond any of the telescopes on Mauna Kea. A planet at the distance of Jupiter, however, would require a resolution of only .5 arc second, which is possible with the Keck telescope. In fact, with adaptive optics, the resolution of the Keck telescope should be as low as 0.2 arc second.

Most of the early effort will be directed toward indirect detection—both astrometric and spectroscopic studies. Later, however, considerable attention will be paid to direct detection. Interferometry, as we've seen, will be particularly important during this phase. The two Keck telescopes are 85 meters apart and when hooked up as an interferometer they will act as two sections of a mirror 85 meters in diameter. With just the two telescopes we will get much better resolution, but only part of the image. For a more complete image other telescopes will be needed along with them.

Plans are now under way for the construction of four small "outrigger" telescopes that can be moved. They will have diameters from 1.5 to 2 meters. Used in conjunction with the two Keck telescopes, they will give much better coverage of the image. According to current plans they will be moved between 18 different fixed positions. Signals from the larger telescopes along with the four outriggers will be routed to the basement of the Keck II telescope where they will be combined.

With such a system, along with adaptive optics and the best seeing in the world at Mauna Kea, the chances of finding a planetary system around some of the nearby stars are promising.

CHAPTER 14

Searching for the Origin of the Solar System

If we are to recognize other planetary systems, particularly systems that are just forming, we must understand how our own system formed, and how it evolved. Astronomers have developed models of its origin, but reaching back into the past, trying to reconstruct events that happened so long ago with so few clues is difficult. We don't have much to work with. Most of the objects around us now, the planets and moons, have changed dramatically since they formed. The model is therefore incomplete in many ways.

According to the model the solar system came into being about 5 billion years ago. The universe was more violent then than it is today because many of the stars were giants, and giants live short lives that end in violence—as supernovae. Supernovae are, without a doubt, the most spectacular events in the universe, but they also serve a useful purpose: they supply the ingredients for new stars. And one such supernova supplied the ingredients for our sun and its family of planets. It supplied a giant nebula—the solar nebula—a gaseous cloud composed mostly of hydrogen with some helium, and small amounts of other elements. At first this cloud was shapeless, looking perhaps like the giant gaseous blob in the constellation Orion. It was huge relative to our present-day solar system, but like everything in the universe, it was under the influence of gravity. Over thousands of years, self-gravity pulled it inward, and gradually it became more and more spherical. Initially, its spin was small, but as it shrunk, its spin increased, just as the spin of a skater increases when she pulls in her outstretched arms.

For thousands of years matter continued falling from all directions toward the center of mass of the cloud, but as its spin increased, an outward force developed along the plane of the spin. This force kept the particles in this plane from falling, but it had no effect on the infall elsewhere. Particles from all other directions therefore continued their inward flight until finally the giant sphere started to flatten. It became disklike, with a dense bulge at the center. Most of the matter was now in this bulge, and it was therefore much denser than the disk surrounding it. At the center it was particularly dense because of the weight of the layers above it. These layers compressed it, generating heat, and as the infall continued, temperatures skyrocketed. This central bulge would eventually become our sun.

Radiation, generated by the hot gas at the center, soon began to flood out through the dense layers of the cloud, but it could not penetrate them easily, and as a result the temperature dropped off in the outer regions of the solar nebula. It developed what is called a temperature gradient.

Although the nebula was composed mostly of hydrogen and helium, a small amount of it (1 percent) consisted of atoms of heavier elements, and these heavier elements eventually began to condense out of the cloud. Temperatures just outside the central bulge, where Mercury is now, were high, and heavy elements such as iron, nickel, and silicon compounds of magnesium condensed here. Further out, near the present position of the Earth, temperatures were lower, and silicates of iron and magnesium and various oxides condensed. Finally, out near Jupiter, carbon, nitrogen, oxygen, and ammonia condensed, along with water. This is generally the distribution we find today: the inner planets have dense cores composed mainly of heavy elements while most of the lighter elements are in the gas giants in the outer regions of the solar system.

The elements "popped out" of the nebula as tiny grains, and were soon forced, as a result of gravity, into a sheet along its mid-plane. This sheet was, in many ways, like Saturn's rings, but much larger, and unlike Saturn's rings, it was immersed in a dense fog.

Close to the sun the velocity of the grains was high, but it dropped off as you moved outward, just as the orbital speeds of the planets drop off as you move outward in the solar system—a consequence of Kepler's laws of planetary motion. Because of this differential velocity, grains began to strike one another, coalescing to form larger grains; then gravitational instabilities began to develop in the disk, and it began to break up. Clumping and aggregation of these grains continued, and they grew in size. Soon they resembled rocks, with some growing to a kilometer and more in diameter.

The solar system was now about 80 million years old. The rocks, or planetesimals, as they are now called, continued to collide and coalesce. Although they had high orbital velocities, their velocities relative to one another were low, and they coalesced when they collided, rather than break apart.

Soon the first protoplanets—the forerunners of our present-day planets—appeared. Four were close to the hot central bulge, and four were widely separated farther out. At this stage they were little more than rocky spheres surrounded by huge, dense atmospheres of hydrogen and helium. In addition, the overall system was still immersed in a huge gas cloud.

As the protoplanets were forming, the protosun at the center was becoming increasingly dense. It's core now had a temperature of several million degrees, and it was continuing to rise. When it reached 15 million degrees, nuclear reactions were triggered and an intense wind—the "solar gale"—rushed out through the solar system blowing the fog from the newly forming planets. The gale stripped the atmospheres completely from the inner planets, but it was not powerful enough to blow the hydrogen and helium from the outer planets. More massive, these planets had stronger gravitational fields, and they therefore retained their atmospheres. They still have them today.

The inner planets were now barren, desolate rocks with no atmosphere. Internal heating, however, would eventually produce volcanism, which would give them new atmospheres.

Looking around today we see that there is more to the solar system than the sun, the planets, and their moons. All of the debris of the solar nebula didn't go into planets and moons. Much of it, in fact, is still in a belt between Mars and Jupiter. The debris here is in the form of rocks that astronomers refer to as asteroids; the belt itself is called the asteroid belt. Also, in the outer reaches of the solar system are large numbers of comets—icy spheres that develop long gaseous tails when they approach the sun.

Our knowledge of the early solar system is still quite limited. We have only a hazy idea of what happened, and it is, unfortunately, a difficult era to study because of the evolution that has occurred. The only objects left unchanged are the asteroids and comets; they are therefore an invaluable resource. Several people at the Institute for Astronomy are, in fact, studying comets, hoping to learn more about the early solar system from them. Dave Jewitt of the Institute for Astronomy feels there is still much to be learned from them. "When you first read the literature you get the impression that a lot is known about comets," he said. "People say, 'This is true . . . that is true.' But, in fact, very little is known for sure."

COMETS

With long glowing tails and a ghostlike aura, comets struck fear into the hearts of the Ancients. They were looked upon as omens of disaster. Much of this fear was dispelled when Edmund Halley of England showed that many of them were periodic. Using Newton's law of gravity he calculated the orbit of a particularly bright one that had appeared several times in the past. According to his calculations it had a period of 76 years, and would return again in 1758. And indeed it did, but by then Halley had been dead for many years.

Comet Halley in 1986. (Courtesy Scott Johnson and Mike Joner)

When comets are a long distance from the sun, they look like tiny fuzzy stars in the telescope. As they get closer, however, they develop a coma—a gaseous envelope that extends out sometimes as far as 100,000 miles. Finally, a tail develops that stretches through space for a million miles or more. This tail is always directed away from the direction of the sun. In fact, if you look closely, you will see that there are actually two tails that have slightly different directions. Only one, the smoothest of the two, is pointed directly away from the direction of the sun; it is called the gas tail. The other, a more lumpy tail composed mostly of dirt particles, is called the dust tail.

Why do tails form, and in particular, why are there two tails? The answer came in 1950 when Fred Whipple of Harvard University put forward his "dirty snowball" model. According to this model, comets are made up of a conglomeration of frozen materials, including water, methane, ammonia, carbon dioxide, carbon monoxide, and hydrogen cyanide. The comet's tail is produced when the sun evaporates the outer layers of this frozen snowball. The ices sublimate, forming a gas cloud around the nucleus. As the gases pour out from the dirty snowball, they carry the dirt and debris in it with them, so the cloud is laced with tiny dirt particles. As the comet gets closer to the sun and a tail forms, the solar wind pushes it away from the direction of the sun. But the gas is much lighter than the debris with it, and it is easily pushed. The dirt particles, being heavier, form a separate tail in a slightly different direction.

In addition to the problem of the constitution of comets, there was also the problem of their periodicity. Halley had shown that some comets were periodic, but eventually it was discovered that all comets did not seem to be periodic; every year or so a comet would appear that had never been seen before. Astronomers soon realized that there were two types of comets: long-period and short-period. The long-period comets had periods of hundreds of thousands of years; the short-period ones had periods that ranged from a few years up to a maximum of about 200 years.

Why was there such a tremendous difference? Jan Oort of

Holland provided an answer at about the same time Whipple published his dirty snowball model. Oort hypothesized that the solar system was surrounded by a dense swarm of comet nuclei—billions of dirty snowballs in a shell that extended from 50,000 astronomical units (AU) out to 150,000 AU from the sun. He assumed that they had been formed somewhere near the asteroid belt, and had been expelled to this distant shell by Jupiter.

According to Oort, a passing star would occasionally perturb one of the comets within this cloud, and it would begin a long journey toward the inner solar system. Tracing out an elliptical—cigar-shaped—orbit, it would take 100,000 years to reach the planets. If unperturbed in the inner solar system, it would continue out to the Oort cloud, pass through it, and appear again at the inner solar system 100,000 years later. Occasionally, however, one of these comets would be perturbed by Jupiter or Saturn while it was in the inner solar system, and would take up a new orbit in this

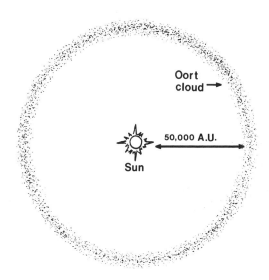

Schematic of the Oort cloud.

region—much less elongated, and of much shorter period. It would become a short-period comet.

Interestingly, at about the time Oort put forward his hypothesis, another one was suggested by Gerard Kuiper of Yerkes Observatory in Wisconsin. At the time there was still considerable controversy about the origin of the solar system, but Kuiper was convinced it had formed from a solar nebula. He had, in fact, put forward an evolutionary model only a few years earlier. Kuiper was sure that the solar gale would have vaporized all primeval cometary material in the inner solar system. He believed, however, that there was still enough of this material in the region beyond Uranus's orbit to create a belt of comet nuclei. This belt, according to his calculations, was 35 to 50 AU from the sun.

It was an interesting idea, but few people paid any attention to it, and it was soon forgotten. Strangely, though, 40 years later it was resurrected and vindicated. In 1988, physicist Martin Duncan of Queens University in Kingston and astronomers Thomas Quinn and Scott Tremaine of the University of Toronto set up a

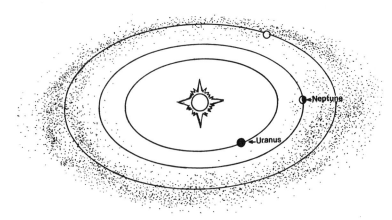

100 AU

Schematic of the Kuiper belt.

computer simulation of the creation of short-period comets, assuming they were comets from the Oort cloud that had been perturbed by Jupiter. To their surprise the computer simulations didn't work; they couldn't get short-period comets similar to those observed.

The trio then tried the same simulations assuming the short-period comets came from the Kuiper belt. Everything worked beautifully—the match to the short-period comets was amazing. The only problem was that no one had ever observed the Kuiper belt.

The short-period comets, it seemed, were coming from the Kuiper belt; the long-period ones, on the other hand, were still assumed to have originated from the Oort cloud. To a number of people, however, it seemed strange that comets existed only in two widely separated belts. In particular, why was there a large comet-free region between them? Jack Hills of the Los Alamos Laboratory in New Mexico had already suggested in 1981 that this region wasn't free of comets. He postulated an "inner comet cloud" that started from Neptune's orbit, which we now assume to be the inner edge of the Kuiper belt and extended out to the Oort cloud. He believed it contained 200 times as many comets as the Oort cloud. Astronomers are now taking his idea more seriously.

OBSERVING THE KUIPER BELT

"My office is cramped," said Dave Jewitt of the Institute for Astronomy as he greeted me at the door. "Let's go out on the balcony."

It was a warm, cloudless day as we sat down around a circular white table. Dense vegetation covered the rolling hills in the distance. Only a couple of months earlier Jewitt and Jane Luu of Harvard University had discovered the second of two objects in the Kuiper belt.

Jewitt was born in North London, England, in what he re-

Dave Jewitt.

ferred to as a "working class area." He went to a school for working class children. "England is divided up into a class structure," he said. Then, shaking his head he added, "It's a strange society, and I felt its effects."

Both of his parents did manual work in factories. Although they had little education they were very supportive of an education for him. "They recognized it as something different . . . something better," said Jewitt. "They were very keen on it."

Jewitt became interested in astronomy when he was about 6 or 7. "At the time I had no idea I would go into astronomy. One of the features of growing up in an economically depressed area is that you don't know what is going on most of the time. Nobody tells you what route you have to take to get educated. I didn't even know what it meant to be educated. I had no idea how you became an astronomer."

His first telescope, a 2-inch refractor, introduced him to the moon and planets. "I was fascinated," he said. He was so fascinated that he began making telescopes, starting with a 4-inch reflector. Over a period of several years he continued building larger and larger telescopes, culminating in one 12½ inches in diameter.

London was not the ideal place to observe the heavens; the sky was bright and it was difficult to see the stars so Jewitt spent most of his time observing the moon and planets. He soon found that he wanted to do more than just observe them; he wanted to use his telescope to determine something about them. So he began making drawings of the markings on Jupiter, and immediately noticed he could use these drawings to determine its period of rotation. At about this time he also became associated with the British Astronomical Society, and began attending their monthly meetings.

He obtained his undergraduate degree from the University College of London, where he majored in astronomy and physics. "They called it a joint astronomy/physics degree," he said, "but the curriculum was mostly physics." In 1979 he came to the California Institute of Technology for graduate work; his thesis, which was done under Jim Westphal, was on comets and planetary nebulae.

During 1992 and 1993 he and Jane Luu identified two objects in the outer solar system. "They are the most distant objects that have been detected in the solar system," said Jewitt. "They're each about 250 kilometers in diameter, and we're sure they are part of the Kuiper belt." He paused. "We think we've found one of the biggest blocks of primeval material in the solar system."

The two objects were discovered using a large CCD on the 88-inch University of Hawaii telescope on Mauna Kea. The procedure they used was simple and straightforward. They would take four pictures of a star field using the CCD, with a fixed time interval between each, then search for objects that appeared to move relative to the stars. For this phase of the work they used a computer that allowed them to blink back and forth between the four pictures. The stars would all remain at the same position; an asteroid or comet, however, would appear to jump as the computer was blinked. They searched, in particular, for very slow moving-objects. As we've seen, objects in orbit around the sun obey Kepler's laws; this means that those nearest the sun move the fastest, and as you go outward, objects move slower and slower. Anything beyond Neptune would therefore move very slowly.

The objects found by Jewitt and Luu were 23rd magnitude, so they're exceedingly dim. This is roughly the limit of most large telescopes not equipped with a CCD. With a CCD, stars up to 28th magnitude can be observed.

Jewitt was particularly excited because the two objects were in one square degree. If you extrapolate this to the entire ecliptic, it implies there should be at least 10,000 objects of this size (250 km) in the Kuiper belt. Furthermore, it is known that small comets are more common than large ones, and comets the size of Halley's, for example, would therefore be much more plentiful than the ones Jewitt and Luu found. Halley's comet is only 10 km in diameter, compared to 250 km for the objects they found. Jewitt estimates that there are one to ten billion objects the size of Halley's comet in the Kuiper belt.

Jewitt elaborated on the discovery, pointing out its importance to me. "It is significant in that it allows us to get close to the origin of the solar system; out where these objects were found the temperatures are low—50 K or less—and not much has happened [since the beginning of the solar system]. Any gases are frozen, and the time between collisions is long. The hope is that we can use the physical observations of objects in the Kuiper belt to understand the early solar system. We hope to find 50 or 60 of these objects over the next few years, then look at their shapes, sizes, rotational properties, colors, surface compositions, and relate these properties back to the formation of the planets. It's a promising area."

The CCD that Jewitt has been using in his research has 4

million pixels. "It's the best available anywhere," he said. "It's coated with an antireflecting material that gives it a quantum efficiency close to 90 percent."

I asked him if he thought it would ever be possible to observe comet belts in extrasolar planetary systems. He hesitated, then said, "It will likely become possible. The submillimeter telescope would be useful in searching for them." Then, with a smile, he added, "If we want to understand them, though, we must first understand the comet belts in our system."

Jewitt has, indeed, been using the submillimeter telescope in his work. One of the problems he has been looking into, using it, is mass loss from comets. Many of the particles that are lost are in the submillimeter range, and are best observed with the Maxwell telescope. "It's a difficult problem," said Jewitt. "Right on the cutting edge of research."

Jewitt was recently able to get an excellent image of comet Shoemaker–Levy. It was discovered near Jupiter on March 24, 1993; two days later he was observing on the 88-inch telescope and recorded a CCD image of it. "It's an intriguing object with 20 distinct nuclei," said Jewitt. "We're seeing dust in the coma of each one. We can't see the bare nuclei, so they may be small—only a few hundred meters across. All 20 are co-moving through space, but we expect them to start spreading out quite soon. We don't know the details of the breakup: were 20 shot out at once, or one at a time? We're not sure. It's probably been a year now since it happened, and it should be visible for quite a while. We think it's rare, but maybe that's because people have never looked for such things before. We plan on monitoring it as long as possible."

SHORT- AND LONG-PERIOD COMETS:
HOW DO THEY DIFFER?

Karen Meech of the Institute for Astronomy is also working on comets. Her major interest at the present time is determining

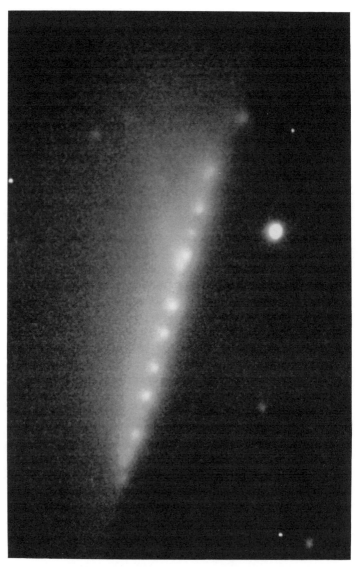

Comet Shoemaker-Levy near Jupiter. (Courtesy Dave Jewitt and Jane Luu)

the physical differences between the short-period and the long-period comets. "For years people have thought that these two groups are physically different," said Meech. "Comets that have spent their entire lives at the very low temperatures of the Oort cloud should contain a greater abundance of highly volatile material than the short-period comets in the inner solar system. I'm hoping that the observed differences between the two groups will help us learn more about how the solar system formed, and what conditions were like at that time."

Meech was born in Denver, Colorado, and lived there until

Karen Meech.

she went to college. Throughout elementary and high school she
wanted to become an astronaut, but settled on astronomy in college. She obtained an undergraduate degree from Rice in Texas.

When she graduated she decided to take a year off to work in
astronomy before going to graduate school. She and a friend split
a job at the American Institute of Variable Star Observers in Boston. Professional astronomers run the organization, she said, but
all observations are made by amateur astronomers; they monitor
variable stars and send in reports.

With only a half-time job she had a lot of time on her hands,
so she went to MIT to see if there were any part-time openings. Jim
Elliot of the Astronomy Department gave her a half-time job working on planetary rings. After she had worked for him for a while
he asked her if she would be interested in coming to MIT for her
graduate degree. "I thought it was the ugliest place in the world,"
she said. "But I went, anyway." She got her degree in planetary
science; her thesis was on the evolution and activity of Halley's
comet, and a comparison of Halley's comet to more distant comets.

Meech has now looked at about 50 comets, both short- and
long-period, to see how bright and how active they are, when they
began to get active, and when they turned off their activity. Her
objective was to determine differences in the two groups, and she
has, indeed, found several. One of the first things she noticed was
that the long-period comets are much brighter in general than the
short-period ones. She also noticed that they "turned on," in other
words, started to produce a coma, much farther from the sun; this
indicated they had more volatile materials than the long-period
comets. She mentioned, however, that Halley's comet is an exception; it turns on relatively early.

"There's still some controversy whether this difference is due
to aging, or whether the long-period comets are just bigger," said
Meech. "The most logical explanation is that we're seeing some
evolution as the comets grow older. You would expect a buildup
of dusty layers on a comet as it gets older, and the ices leave, and
you would also expect volatile ices to disappear."

Most of the work done so far has been done using a 1024- by 1024-pixel CCD. Meech mentioned, however, that a group at the institute is working on an even larger CCD: 8192 by 8192 pixels. She expects to use it as soon as it is finished. She said it would give excellent sky coverage—almost a square degree.

Meech was codiscoverer of a flare-up in Halley's comet in February of 1991. Halley faded as expected as it receded from the sun, and when it was about 10 AU from the sun it completely turned off. Then suddenly at 14 AU it brightened from magnitude 24 up to magnitude 16. This was a huge increase, and was quite unexpected. What was particularly surprising is that energy from the sun is generally thought to be responsible for triggering such outbursts, and Halley was a tremendous distance from the sun when it happened.

"From the extension of the coma we could tell what the ejection velocity was," said Meech. "Also, the extent of the brightening tells you something about the volatile materials. As to what caused it, there's been a lot of speculation. There are many possibilities—impact of a giant boulder, a solar flare shock wave. But the most likely explanation is volatile materials under the surface. You sometimes get a pocket of gas that builds up a tremendous pressure, then outbursts. This is what it looked like."

CHIRON

Both Meech and Jewitt have worked on an object called Chiron, discovered in 1977 by Charles Kowal using a blink comparator. It didn't look like a comet to Kowal, but at 200 to 800 km in diameter it was relatively large for an asteroid.

Brian Marsden of Harvard calculated its orbit and found it was more circular than that of asteroids and comets. Most of its orbit was between Saturn and Uranus, but a small section was inside Saturn's orbit. Its period was 50.7 years.

It was assumed initially to be an asteroid, but its orbit was farther out than most asteroids. Furthermore, in late 1987 it suddenly brightened—something asteroids do not do. The brightening indicated it might be a comet, but there was no evidence of a coma. In April 1989, however, Meech and Michael Belton of the National Optical Astronomical Observatory in Tuscon showed that it did have a coma. The discovery was verified in November by H. Spinrad and M. Dickinson.

Jewitt observed the object with the James Clerk Maxwell submillimeter telescope in late 1991, hoping to obtain an accurate measure of its diameter. He determined it was about 300 km, making it the largest comet nucleus ever detected.

As Jewitt and others continue to study comets and other primitive objects in the solar system, valuable insights into the origin of the solar system will no doubt be uncovered.

CHAPTER 15

The Future: Other Telescopes

Despite the large number of telescopes available to astronomers at the present time there is still considerable pressure for telescope time. Roughly two-thirds of all requests for time go unfulfilled, and the competition for time is likely to get much worse in the next few years unless more telescopes are built. And many are, indeed, being built, several on the summit of Mauna Kea.

GEMINI

In 1986 the National Optical Astronomical Observatory (NOAO) and the Association of Universities for Research in Astronomy (AURA) met to consider the future of nighttime astronomy in the United States. They were concerned with the shortage of telescopes, and the difficulties many astronomers were having getting telescope time. How could the problem be alleviated? They decided that two large telescopes were needed, one for the northern hemisphere and one for the southern. Both should be approximately 8 meters in diameter, with one to be placed on Mauna Kea and the other in Chile. In September 1989, a proposal was delivered to the National Science Foundation for the funding of the two telescopes. The project was to be called Gemini.

Scientists and engineers at Steward Observatory of the University of Arizona had been working for years on the technology for developing large mirrors, and it was assumed they would

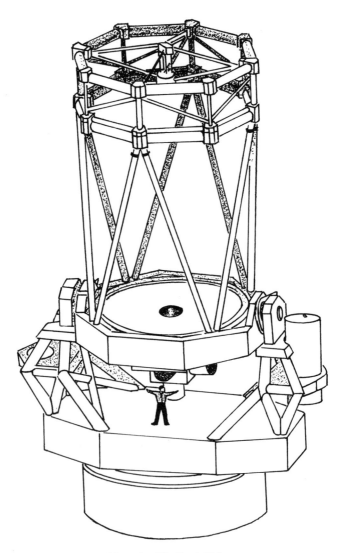

Schematic of the Gemini telescope.

supply the mirrors. Roger Angel and his team at Steward had developed an innovative method of casting blanks with concave surfaces by spinning the disk while it was molten. Since it usually takes several years to grind the surface to the required concave shape, this innovation would give considerable savings in both time and cost. Angel had also developed a technique for casting disks with a honeycomb structure beneath the mirror surface to give it extra strength. Using borosilicate glass, he and his team had cast three 3.5-meter disks. One of the major reasons why it was assumed they would supply the blanks was that they had been supported for several years by the National Science Foundation (NSF), and the NSF was funding the two Gemini telescopes.

But Corning Glass of New York also had considerable experience in large disk technology. Many years earlier they had cast the blank for the 200-inch Palomar mirror. Furthermore they had developed a new type of glass with an extremely low expansion coefficient that they called ULE (Ultra Low Expansion). Its expansion coefficient was considerably lower than that of borosilicate. And they were able to cast relatively thin meniscus disks that were flexible, yet rigid enough to withstand considerable distortion.

A cost estimate of the two projects was drawn up and it was determined that they would cost $176 million. The NSF agreed to come up with half, namely $88 million, which they stressed was

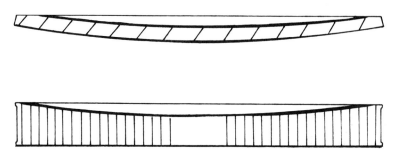

Upper: A thin-meniscus lens similar to that cast by Corning. Lower: Lens with honeycomb structure beneath it as cast by Angel and his group.

the limit—there would be no overruns. The additional money would have to come from partners in the project. A number of astronomers expressed concern; this would mean they would have to share the telescopes, and would only get half of the total telescope time. The United Kingdom and Canada both agreed to come in as partners, with the United Kingdom taking 25 percent and Canada 15 percent. This left 10 percent, and since one of the telescopes was going to be placed in South America, several countries in South America were approached. Chile agreed to pick up 5 percent and Brazil and Argentina the remainder.

With the plans complete, a meeting was held in September 1992 to select a vendor for the mirror blanks. Everyone expected it to go to the University of Arizona but to the surprise of many it was awarded to Corning.

Sidney Wolff, the acting director of the project, justified the selection by stating that the Corning disk was the one most in line with Gemini's specifications. The committee, she said, was convinced it was the one most likely to succeed; furthermore, Corning's bid was considerably lower than the University of Arizona's. One of the major reasons for the selection, however, was that the University of Arizona's bid stated that AURA would have to share in the risk of a failed casting, which would be about $3 million per disk. Corning didn't make similar demands because ULE glass, unlike borosilicate (molded into a honeycomb), can be reheated to take out small flaws, and if the entire casting is a failure, the material can be reused. This is not the case with borosilicate glass.

Many astronomers were upset by the AURA decision. Considerable money had been put into the University of Arizona project by the NSF over the years, and its future depended, to some degree, on getting the Gemini contract. The backlash was so great, in fact, that within two months of the announcement the NSF asked for an independent appraisal. As a result a committee of seven American astronomers, chaired by James Houck of Cornell, was set up. Strangely, there was no representation from the foreign partners, nor any experts on glass casting on the committee. Within a few months the new committee arrived at a decision, which they reported to the NSF: the Corning meniscus disk was risky. The University of Arizona had more experience than Corning; it had just cast a 6.5-meter disk, while Corning's largest was 4 meters. The committee strongly urged AURA to revert to the borosilicate glass disk.

AURA, however, ignored the recommendation, and has proceeded with their plan for using a meniscus disk. But Congress and the NSF have the last word, and they have called for further study. Because of the controversy the telescope has been redesigned so it can accept either disk. At the present time no one is sure which it will be.

Despite the slightly higher cost of the University of Arizona's disk, there is an advantage in purchasing the disk from them. The University of Arizona, unlike Corning, is capable of producing a finished product; in other words, it can also do the figuring and final polishing of the surface. Corning cannot, and if the Corning disk is purchased, another vendor would have to be found to do this. Worldwide, there are less than half a dozen companies that can figure and polish surfaces this large. The expected cost of doing this is about $5 million for the two mirrors.

The meniscus and borosilicate disks are both so thin that they will have to be supported by an elaborate active system, much in the same way the Keck mirrors are. The technology for doing this with an 8-meter mirror has not yet been completely developed. At present a system of 172 hydraulic activators is planned. The meniscus lens has a slight advantage in this regard in that activators and sensors can be placed anywhere on the glass; on the borsilicate disk they can only be placed in the honeycombs.

As we mentioned earlier, an active system must correct for two things: the effect of gravity on the mirror as it is moved from position to position, and distortion caused by the wind. Both of these effects would be more significant on a large mirror blank such as this than they would be on the smaller Keck mirrors. The system will therefore be relatively complex.

Although both mirrors would have to be supported by an active system, the borosilicate mirror has an advantage. With its honeycomb structure it is stiff enough to withstand most distortions caused by the wind. Little, if any, correction will be required for it.

What is likely to happen as a result of the controversy? There is the danger that Congress will get fed up and decide not to fund the project. With the Hubble disaster and the failure of the Mars mission, there is little sympathy for further failures. If the controversy is not resolved there is also the possibility that the United States will go ahead on its own, and build one of the telescopes. It would be placed on Mauna Kea. This alternative has been looked into, but there is a difficulty: one telescope costs more than half of the cost of two telescopes—about $18 million more. This means that $88 million would not get astronomers the telescope they would like. Many cuts would have to be made in the design, and U.S. astronomers would not be happy.

Despite the problems the project is going forward, and completion of the telescope on Mauna Kea is scheduled for 1998.

PROJECT SUBARU

Several years ago the Japanese selected Mauna Kea for their national telescope. It is called Subaru after the Japanese word for the star cluster Pleiades. Work on the site—soil testing and leveling—began in 1992, and in mid-1993 the massive concrete pier that will support the telescope structure was poured.

Japan has decided to use Corning's ULE glass, and Corning is presently working an extremely thin 8 meter blank for them. It will only be 20 centimeters thick, which will make it flexible enough so that its shape can easily be adjusted with a computer-controlled active support system.

Corning is fabricating the blank by taking 40 hexagonal sections, each about 1.5 meters in diameter, and fusing them into a disk 8.3 meters in diameter. They do not spin their blanks, but get the same effect, a slight concave surface, by reheating the disk in a preshaped mold. It will take three years to produce the blank,

Preparing the base for the Subaru telescope.

Schematic of the Subaru dome. (Courtesy Subaru Observatory)

Schematic of the Subaru telescope. (Courtesy Subaru Observatory)

and three more years of grinding and polishing to complete the mirror surface.

The active support system has been designed and will consist of 264 activators and sensors. The mount for the telescope will not be an equatorial mount (with one axis pointed to the celestial pole) as is the case in most telescopes. Rather, it will be what is called an altitude–azimuthal mount—a simple up–down, right–left axis. A mount of this type can be used because highly accurate control of the telescope's motion is now possible via computers. With their drive system, the Japanese hope to track with an accuracy of 0.1 arc second.

They have also put a considerable amount of work into the design of the observatory building, using wind tunnel modeling to determine the airflow patterns through the observatory. As a result of this testing they have decided to build a cylindrical dome, rather than the traditional semispherical dome.

The telescope will be sensitive to both the infrared and the visible, and will be equipped with an impressive array of instruments, including large infrared detectors and CCDs. One of the CCDs will be composed of 64 arrays of 1000 by 1000 pixels giving it an extremely wide field of view.

The telescope will be controlled from a building adjacent to the observatory. Headquarters for the observatory will be built in Hilo, adjacent to the Joint Astronomy Centre.

THE SMITHSONIAN SUBMILLIMETER ARRAY

Construction has also begun on an array of submillimeter telescopes that will be located just north of the James Clerk Maxwell telescope in Submillimeter Valley. It will consist of six 20-foot (6 meter) parabolic antennas that will be placed on concrete pads. In all there will be 24 pads, so many configurations of the six antennas will be possible.

The antennas have been designed to be transportable using a specially designed 16-foot-wide carrier with large rubber tires.

One of the dishes that will be used in the Smithsonian Array. (Courtesy Smithsonian Institution Astrophysical Observatory)

The carrier is capable of easily moving the antennas, even over relatively rough terrain.

Each of the antennas will be connected to a computer that will combine the signals, making the array equivalent to an instrument 1500 meters across. The array will be assembled and tested in Westfield, Massachusetts, before it is brought to Mauna Kea.

A two-story control building containing an electronics lab, a shop, offices, and lounge will be built. Up to ten employees will be at the facility at one time. The project, which will be funded by the

Radio telescope of VLBA.

Smithsonian Institution, will cost $45 million. Its planned completion date is 1997.

What will astronomers search for using the array? Submillimeter radiation, as we have seen, penetrates clouds of dust that block visible light. One of the major projects, therefore, will be the core of our galaxy, which is largely obscured in the visible region of the spectrum. Using it, astronomers can measure the velocity of stars, dust, and gas in this region, and may be able to tell us for sure whether a massive black hole is hidden there. It will also be used to look at the cores of other galaxies.

THE VERY LONG BASELINE ARRAY

Finally, one leg of the Very Long Baseline Array (VLBA) radio telescope was recently completed on Mauna Kea. This array will give astronomers an unprecedented look at the radio universe using a technique called very long baseline interferometry (VLBI). The array consists of ten radio telescopes, each 25 meters in diameter. They are located at various sites across the continental United States, including Mauna Kea and the Virgin Islands.

The dishes will be controlled from an operations center at Socorro, New Mexico, via telephone lines. A computer will operate the entire array—monitoring the telescopes at each site. Remote control will allow all telescopes to be pointed at the same point of the sky at the same time. Data will be taken on magnetic tapes and the tapes will be shipped to Socorro for correlating and transforming into maps of the radio sky.

Epilogue

We have explored the history of the Mauna Kea observatories—how they began and developed into the largest observatory complex in the world. And with work starting on the Japanese telescope, Gemini, and the Smithsonian Array, it will continue to grow for many years. We have also looked at some of the research that is going on there, research on black holes, cosmology, galaxies, quasars, clusters, stars, and closer to home, research on comets and the search for extraterrestrial life.

Not only are more telescopes being built, but the instruments used with the telescopes are being significantly improved. Larger and larger solid-state arrays—CCDs, infrared and submillimeter detectors—are being constructed, along with more powerful and efficient spectrometers. All of this will have a serious impact on astronomy.

New techniques are being developed. Some of them will be used in conjunction with the Keck telescope. When the twin to Keck I is complete the two telescopes will be used together via interferometry; this will greatly increase their effective size. Interferometry will no doubt be used extensively in the future. Plans are already under way to use it with the Very Large Telescope Array (four 8-meter telescopes) that is being built in Chile. Furthermore most telescopes now being built will be equipped with adaptive optics, vastly increasing their resolving power.

One of the major changes that will be coming over the next few years, however, will be remote observing. Astronomers coming to Hawaii to use the Keck telescope, for example, will not go up the mountain to the observatory; they will do their observing at Keck headquarters in Waimea. Some remote observing is now being done by UKIRT and CFHT. Remote observing will no doubt be used increasingly in the future, and with it will come a significant change in the role of astronomers. They will no longer travel to observatories, but will stay in their office, sitting in front of computer screens that are linked to the observatory. Data will come to them directly from the telescope, and they will be able to analyze it using software in their computer.

There is still much room on the summit of Mauna Kea for other telescopes so it will likely continue to grow for many years. Unlike most other observatories, conditions at the summit are not likely to deteriorate with the encroachment of man. The island has a small population, so light pollution is minimal, and it will likely stay that way. And with little industry the air will no doubt remain highly transparent and clear of dust and smog.

As we enter the 21st century a new age of astronomy is dawning, with some of the universe's most perplexing mysteries on the verge of resolution. And as they are resolved, even deeper mysteries will no doubt be uncovered. With the advances in telescope technology and innovative techniques now being used on Mauna Kea, the future of astronomy never looked brighter.

Glossary

Absorption line   Dark spectral line superimposed on bright background.

Acclimatize   To adapt to high altitude by temporary layover.

Accretion disk   Flattened disk of matter whirling around a star or black hole.

Active galaxy   A galaxy that is emitting large amounts of energy from near its core.

Actuator   A device beneath a mirror that pushes on the mirror surface to change its curvature.

Adaptive optics   A technique for increasing resolution by correcting for the turbulence in the atmosphere.

Altitude   Angular distance above or below the horizon, measured along a vertical circle to the celestial object.

Altitude–azimuthal mount   Mount for a telescope that drives along altitude and azimuthal axes.

Aluminize   To put a thin reflecting coating of aluminum on a mirror.

Azimuthal axis   Angle along celestial horizon, measured eastward from the north point to the intersection of the horizon with a vertical circle passing through the celestial object.

Binary system   A double star system, or system of two celestial objects.

Black hole   A region of space-time from which nothing, not even light, can emerge.

Blazar   A quasar seen along line of sight of emerging jet.

Blink comparator   A device for blinking back and forth between two photographs or pictures of the sky, usually taken at different times.

Blue straggler   A large blue star in a cluster that should have evolved to be much redder, but for reasons not yet fully understood is bluer and brighter than it should be.

Capacitance   Power of an electrical device for storing static electrical charge.

Capacitor   A device for storing static electrical charge.

Cassegrain focus   An optical arrangement in a reflecting telescope in which the light rays reflect from a small secondary mirror focus behind the primary mirror.

CCD (charge-coupled device)   A solid-state device designed for the detection of light photons.

Center of mass   Point at which two objects would balance if placed on the ends of a rod.

Centripetal force   Force toward center along radius.

Cepheid   A star that changes periodically in brightness.

Classical physics   Any nonquantum theory. Newtonian physics and general relativity are both classical theories.

Cluster   A group of stars or galaxies.

COBE   Short for Cosmic Background Explorer, a satellite launched in 1989 that measures properties of the background radiation.

Constellation   A group of stars that appears from Earth to be close to one another.

Convection   Heat transfer by mass motion of medium (e.g., air).

Coronograph: A device for creating an artificial eclipse using a small circle along the optic axis of the telescope.

Cosmic background radiation   Background radiation in the universe left over from the big bang.

Cosmic ray   High-energy particles in space that strike our atmosphere producing showers of particles and radiation.

Cosmic string   Hypothetical string in the early universe. May be responsible for structure.

Cosmology   A study of the structure and evolution of the universe.

Coude spectrograph   A spectrograph attached to the telescope at a point far away from the working parts of the telescope. Light is directed to it via a series of mirrors.

Cryostat   A low-temperature receptacle.

Dark matter   Matter astronomers cannot see but know exists in the universe.

Differential velocity   Velocities that differ from point to point.

Doppler shift   The apparent change in wavelength of light caused by relative motion between source and observer.

Double-lobed radio source   Radio source that has energetic regions on either side of it.

Einstein Cross   A system of four closely spaced celestial objects—multiple images produced by gravitational lensing.

Einstein–Podolsky–Rosen paradox   Paradox related to whether you can measure both position and momentum at the same time, in contradiction to the uncertainty principle.

Elliptical galaxy   A galaxy that has an elliptical shape. Usually contains old stars.

Equatorial mount   Telescope mount with the axis parallel to the Earth's axis, so that the motion of the telescope about the axis compensates for the Earth's rotation.

Event horizon   Surface of a black hole. A one-way surface.

Extrasolar   Outside the solar system.

Focal length   Distance from a lens or mirror to the point where converging light rays meet.

Frequency   The number of wave crests or troughs that cross a given point per unit time.

F-type star   A spectral type of star, slightly hotter than our sun.

Galaxy   A large assemblage of stars.

Gamma ray   Highly energetic radiation. Highest energy of electromagnetic spectrum.

Gravitational lens   Deflection of light by a gravitating object.

Great Attractor   A large concentration of superclusters. Attracts many of the superclusters around it.

G-type star   Spectral type of star similar to our sun.

Helium flash   The explosive beginning of helium burning in the dense core of a red giant star.

Herbig Haro object   Young star with strong stellar wind. Clumps of gas strike ambient gas of star causing shock waves.

Herbig star   Young hot star of spectral type A or B that exhibits bright spectral lines.

Hexagon   A six-sided object.

HR diagram   A plot of luminosity or brightness versus surface temperature of star.

Hubble constant (H)   Given by the slope of the line in a Hubble plot (i.e., plot of redshift of galaxies versus their distance).

Hyperbola   One of the family of basic conic curves. Similar to a parabola.

Infrared   Region of the electromagnetic spectrum. Radiation with a slightly longer wavelength than light.

Infrared array   A solid-state detector capable of detecting infrared radiation.

Infrared telescope   A telescope designed to detect infrared radiation.

Interferogram   A picture of interference lines. Interference is caused by two side-by-side beams of light interacting.

Interferometry   The interference of two side-by-side beams of light. Causes alternating regions of light and dark.

Ionization   The process of giving an atom an electrical charge.

IRAS   A satellite launched in 1983 and designed to detect infrared objects in the universe.

Isothermal   At constant or uniform temperature.

Kuiper belt   A hypothetical belt of comet nuclei at outer edge of the solar system.

Laser   Coherent light beam. Waves are in phase.

Light pollution   Stray light affecting observing.

Light-year   A measure of distance. The distance a light beam travels in one year.

Local Group   The group of approximately 25 galaxies that includes the Milky Way galaxy.

Local (Virgo) supercluster   The group of clusters that includes the Local cluster.

Magnetic star   Star with strong magnetic field.

Magnitude (of star)   A measure of brightness. A scale that extends from negative numbers through zero to 28. The smaller the number, the brighter the star.

Malmquist bias   The tendancy when dealing with a field of galaxies to select galaxies brighter than the average.

Nebula   A cloud of interstellar gas and dust.

Neutron star   A star made up of neutrons. Usually only a few miles across.

Noise   A natural fluctuation in a signal.

Nuclear fusion   The joining together of atomic nuclei to form nuclei of greater mass. Energy is given off in fusion of light elements.

Oort cloud   A hypothetical cloud of comets in a shell about 1 light-year from our sun. Presumed source of long-period comets and possibly others.

Parabola   One of the basic conic curves. Formed by cutting a cone parallel to one of the sides.

Peculiar velocity   Velocity resulting from the gravitational attraction of a nearby galaxy, or cluster, or supercluster.

Perturbation (orbital)   A small deviation from the expected orbit.

Photometer   An instrument for measuring the amount of light.

Photon   Particle of the electromagnetic field.

Pixel   Short for picture element. An element of a CCD chip.

Planetesimal   Small, asteroid-sized objects that formed out of the solar nebula. Combined to form protoplanets.

Polarization   A process in which electromagnetic waves with plane of oscillation in all but one direction are removed from a light beam.

Primary mirror   The main large mirror of a telescope.

Protogalaxy   Embryonic stage of a young galaxy.

Proton-proton cycle   A sequence of nuclear reactions in which hydrogen atoms fuse to form helium.

Prototype   A model (usually on a smaller scale) of a system that is to be built.

Quantum mechanics   A branch of physics dealing with the behavior of atoms and their interaction with light.

Quasar   A starlike object with very large redshift. Believed to be colliding galaxies in early universe, yet the cause of it is not fully understood.

Radial velocity   Velocity along the line of sight.

Radiation   Refers to electromagnetic energy. Photons.

Radioactive decay   The process in which atomic nuclei decompose by spontaneously emitting particles.

Radio source   A source of radio waves. Usually a galaxy or quasar.

Radio wave   Electromagnetic radiation with long wavelength.

Red giant   A large, cool star of high luminosity.

Reflector   A telescope in which the primary optical component is a concave mirror.

Refractor   A telescope in which the primary optical component is a lens.

Resolution   A measure of the ability of a telescope to resolve fine detail in the field of view.

Secondary mirror   A small mirror that reflects light from the primary to the eyepiece.

Seeing   A measure of the stability of the atmosphere.

Segmented mirror   A mirror made up of several sections, or components.

Sensor   A device for checking the alignment of the telescope mirror.

Singularity   A point of infinite density. A point where the laws of physics break down.

Software   Refers to programs needed to run a computer.

Solar gale   An explosive event that occurred in the early solar system when nuclear reactions were triggered in the sun.

Solar nebula   Gaseous cloud out of which the solar system formed.

Solar telescope   Telescope designed to use in the study of the sun.

Spectrograph   A device for photographing a spectrum.

Spectrum   A series of bright or dark lines that gives considerable information about a star or other celestial object. Obtained using a spectroscope.

Spiral galaxy   A flattened, rotating galaxy with spiral arms.

Starburst galaxy   A galaxy in which a large number of new stars are forming.

Sublimation   To go directly from the solid state (e.g., ice) to the gaseous state (e.g., steam).

Submillimeter radiation   Radiation with a wavelength slightly less than 1 millimeter.

Supercluster   A cluster of clusters.

Supergiant star   A large star of very high luminosity.

Supernova   A stellar outburst in which a star suddenly increases its brightness by a million times or more.

Tangential velocity   Velocity perpendicular to the line of sight.

T Tauri star   Young star that shows erratic changes in brightness.

Variable star   A star that changes in brightness.

Wavelength   The distance from one wave crest to the next or a wave trough to the next in a wave.

White dwarf   A low-mass star that has exhausted its thermonuclear fuel and collapsed to approximately the size of Uranus.

Z number   Ratio of velocity to speed of light at nonrelativistic velocities. Correction factor is required at relativistic velocities.

Bibliography

The following is a list of general and technical references for the reader who wishes to learn more about the subject. References marked with an asterisk are of a more technical nature.

CHAPTER 1: Introduction

Cruikshank, D. P., *Mauna Kea* (Honolulu: University of Hawaii, Institute for Astronomy, 1986).

Krisciunas, K., *Astronomical Centers of the World* (London: Cambridge University Press, 1988)

CHAPTER 2: The Early Years

Aspaturian, H., "Life on Mauna Kea: The Fascination of What's Different," *Engineering and Science* (Summer, 1988), 11.

Jefferies, J. T., and Sinton, W. M., "Progress on the Mauna Kea Observatory," *Sky and Telescope* (September, 1968), 140.

Krisciunas, K., *Astronomical Centers of the World* (London: Cambridge University Press, 1988).

Waldrop, M., "Mauna Kea(I): Halfway to Space," *Science* (November, 1981), 27.

Waldrop, M., "Mauna Kea(II): Coming of Age," *Science* (December, 1981), 4.

CHAPTER 3: Expansion and New Telescopes

Humphries, C., "The United Kingdom's Giant Infrared Reflector," *Sky and Telescope* (July, 1978), 22.

Sky and Telescope Staff, "Progress on the CFH Reflector," *Sky and Telescope* (April, 1977), 254.

Smith, G. M., "Progress on NASA's 3-meter Infrared Telescope," *Sky and Telescope,* (July, 1978), 25.

Waldrop. M., "Mauna Kea(II): Coming of Age," *Science* (December, 1981), 4.

CHAPTER 4: The Largest Optical Telescope in the World—Keck

Bunge, R., "Dawn of a New Era: Big Scopes," *Astronomy* (August, 1993), 49.

Faber, S. M., "Large Optical Telescopes—New Views in Space and Time," *Ann. N.Y. Acad. Sci.* 422 (1982), 171.

Goldsmith, D., *The Astronomers* (New York: St. Martin's Press, 1991).

Nelson, J., "The Keck Telescope," *American Scientist* (March-April, 1989), 170.

CHAPTER 5: The Continuing Story of Keck

Baker, C., "Mauna Kea: The Best Seat in the Cosmic Theater," *Hawaii High Tech Journal 5* (1991), 6.

Harris, J., "Seeing a Brave New World," *Astronomy* (August, 1992), 22.

Henbest, N., "The Great Telescope Race," *New Scientist* (October, 1988), 52.

Nelson, J., Faber, S., and Mast, T., *Keck Observatory Report No. 90* (Berkeley: University of California Press, 1985).

CHAPTER 6: Visiting the Top of the World

Cruikshank, D., *Mauna Kea* (Honolulu: University of Hawaii, Institute for Astronomy, 1986).

CHAPTER 9: Monster at the Core

Chaisson, E., "Journey to the Center of the Galaxy," *Astronomy* (August, 1980), 6.

Geballe, T., "The Central Parsec of the Galaxy," Scientific American (July, 1979).

Kaufmann, W. III, *Galaxies and Quasars* (San Francisco: Freeman, 1979).

*Kormendy, J., "Evidence for a Supermassive Black Hole in the Nucleus of M31," *Astrophys. J.* 325 (February, 1988), 128.

*Kormendy, J., "A Critical Review of Stellar-Dynamical Evidence for Black Holes in Galaxy Nuclei," Preprint, Institute for Astronomy, University of Hawaii (1993).

Shipman, H., *Black Holes, Quasars and the Universe* (Boston: Houghton-Mifflin, 1980).

Waldrop, M., "Core of the Milky Way," *Science* (October, 1985), 230.

CHAPTER 10: Surveying the Universe

Bartusiak, M., *Thursday's Universe* (New York: Times Books, 1986).

Cornell, J. (Ed.), *Bubbles, Voids and Bumps in Time: The New Cosmology* (London: Cambridge University Press, 1989).

Parker, B., *The Vindication of the Big Bang* (New York: Plenum Press, 1993).

*Pierce, M., and Tully, B., "Distances to the Virgo and Ursa Major Clusters and the Determination of H," *Astrophys. J.* 330 (July, 1988), 579.

Tully, B., "The Scale and Structure of the Universe," Endeavor, *New Series 14* (1990), 1.

CHAPTER 11: Searching for the Ends of the Universe

*Cowie, L., "Galaxy Formulation and Evolution," *Physica Scripta* T36 (1991), 102.

Overbye, D., *Lonely Hearts of the Cosmos* (New York: HarperCollins, 1991).

Parker, B., *Creation* (New York: Plenum Press, 1988).

Parker, B., *The Vindication of the Big Bang* (New York: Plenum Press, 1993).

Preston, R., *First Light: The Search for the Edge of the Universe* (New York: New American Library, 1987).

CHAPTER 12: Stars and Stellar Debris

Editors of Time–Life Books, *Stars* (Alexandria: Time–Life Books, 1989).
Jastrow, R., *Red Giants and White Dwarfs* (New York: Norton, 1979).
Kippenhahn, R., *100 Billion Suns* (New York: Basic Books, 1983).
Maffei, P., *When the Sun Dies* (Cambridge, Mass.: MIT Press, 1982).

CHAPTER 13: Searching for Other Planetary Systems

Abell, G., "The Search for Life Beyond Earth: A Scientific Update," *Extraterrestrial Intelligence: The First Encounter* (Buffalo: Prometheus Books, 1976).
Burke, B. (Ed.), TOPS: *Toward Other Planetary Systems*, NASA Publications, Solar System Exploration Division (1992).
Goldsmith, D., and Owen, T., *The Search for Life in the Universe* (Menlo Park, Calif.: Benjamin Cummings, 1980).
McDonough, T., *The Search for Extraterrestrial Intelligence* (New York: Wiley, 1987).
Rood, R., and Trefil, J., *Are We Alone?* (New York: Scribner's, 1981).

CHAPTER 14: Searching for the Origin of the Solar System

*Meech, K., and Belton, M., "The Atmosphere of 2060 Chiron," *Astron. J.* 100 (October, 1990), 1323.
Sagan, C., and Druyan, A., Comet (New York: Random House, 1985).

CHAPTER 15: The Future: Other Telescopes

Bunge, R., "Dawn of a New Era: Big Scopes," *Astronomy* (August, 1993), 47.

Index